3/96

GOLD RUSHES OF NORTH AMERICA
AN ILLUSTRATED HISTORY

GOLD RUSHES OF NORTH AMERICA
AN ILLUSTRATED HISTORY

Lionel Martinez

THE WELLFLEET PRESS

WELLFLEET

For Barbara, Joe and Calamity.
And also Tony whose idea this book was in the first place.

Copyright © 1990
Copyright © 1990 THE WELLFLEET PRESS
110 Enterprise Avenue
Secaucus, NJ 07094

All rights reserved.
No part of this book may be reproduced or transmitted in any form or by any means, electronic or mechanical, including photocopying, recording, or by any information storage and retrieval system, without permission in writing from the Publisher.

Publishing Director: Frank Oppel
Editorial Director: Tony Meisel
Design Director: Carmela Pereira
Editor: Theresa Stiles
Composition: Meadowcomp Ltd.
Origination: Regent Publishing Services Limited
Printing: Leefung-Asco Printers Limited

Manufactured in Hong Kong.
ISBN: 1-55521-552-1

978
MAR
1990

Contents

CHAPTER 1

GOLD

Gold. Its chemical symbol is Au. It is one of the earliest metals known to man and one of the most sought after throughout history.

It is extremely inert and therefore under normal circumstances does not easily corrode. Because of its durability, the workmanship of ancient artisans using this metal can sometimes be seen in near perfect condition, several thousand years after being crafted.

Gold can be found in a comparatively pure form in nature. It can recognized by its bright yellow lustrous sheen glittering in a stream, or in veins shining from an outcropping of quartz or as small pebbles trapped in tree roots situated in a rushing creek.

Gold is rare and therefore valuable; it can make one very, very wealthy. The very mention of a gold find can cause a mad desire to possess this gleaming metal and inflict tens of thousands of men and women with the psychological disorder known as gold fever. This greedy madness caused fervid mass migrations in the 19th century–since labeled "the great gold rushes of North America."

THE CHARACTER OF GOLD

In the strictest scientific sense gold is just another metal, even though it has some unusual properties.

Au has an atomic number of 79. Its atomic weight is 196.967. Gold will melt at 1,063 degrees Celsius and boil at 2,970 Celsius and its density compared to water ranges from 19.2 to 19.4, depending upon whether it is measured at 0 degrees Celsius or at 20 degrees Celsius or by x-ray crystallography. Gold also has a hardness of 2.5-3.0 on the Mohs hardness scale, a tensile strength of 19,000 psi, a modulus of elasticity of about 11.6x10 to the 6th power, and a face centered-cubic crystalline structure.

What all these scientific statistics indicates is that gold is a pretty amazing metal. It is 19.3 times heavier than water and twice as heavy as lead. Gold is so dense that a cubic foot weighs over half a ton.

Known as the most malleable and ductile of all metals, scientists have estimated that one ounce of gold could be stretched into a wire more than 40 miles long without breaking. Another way to look at gold's incredible plastic quality is the ability of machinists to work a piece of gold into an ultra thin layer measuring one millionth of inch.

Nobility is another attribute of gold. Chemists define nobility as a substance's resistance to corrosion and chemical interaction under normal circumstances. Unlike other metals, for example, iron and silver, gold does not disintegrate when exposed to oxygen, water, salt or any other naturally occurring materials. It is this property which allows gold coins to survive in sea water for a thousand years with little or no noticeable tar-

▲
This spiral galaxy in Coma Berenices is much like our own Milky Way. If the spiral galaxy in Coma Berenices was the Milky Way then Earth would be found in the lower left hand corner about 2/3 of the way to the edge.

nishing, whereas a silver coin after the same length of time would be nothing more than a rough, black undistinguished round lump.

Nonetheless, there are certain manmade acid and alkaline compounds which will dissolve gold. The best well known of these is aqua regia, a potent combination of hydrochloric and nitric acids. Aqua regia will also dissolve almost any other substance except glass (a is very important exception since this powerful solvent is usually kept in glass containers).

Gold is a good conductor of electricity, silver and copper better. But whereas silver and copper corrode easily, gold does not. Gold has found a valuable place in the high tech world of electronics and computers. It is used wherever deterioration due to sparking, arcing, heat, mechanical wear and rust are a problem.

THE COSMIC CREATION OF GOLD

Gold and all of Earth's heavier elements were created billions of years ago in some remote part of space. Singular among the elements, gold, in its luster and shine, reflects the moment of its cataclysmic birth.

Once upon an unbelievably long time ago gold did not exist. There was no earth or moon, no planets at all. The stars were not the stars we know today. It was by today's standards a very strange place indeed, a very young universe heading toward adolescence. Part of the growing up of this young universe was the death of some of its stars. In these deaths the elements, the stuff of planets first appeared in the universe including the stuff of dreams, gold.

The most recent theories on the origins of the universe state that everything began with a big bang. In the smallest part of a second the universe came into being from nothing. Not only did this include energy and matter but time and space as well. As the scenario goes, the universe at the moment of creation was a tiny, hellishly hot dot, about the size of a period on this page, made up of basic atomic particles growing in all directions at fantastic speeds. As this primordial mass expanded it cooled and basic particles began to form larger particles. These larger particles were mere fractions of atoms though atoms are the stuff that elements are made of. In one second the universe cooled to a tepid one billion degrees and all the helium and hydrogen nuclei found in the universe had been created. Each nucleus can be thought of as a submicroscopic sun at the core of a given atom. These tiny suns had no planet-like particles whizzing about them at this time. We still needed electrons to make the atoms complete.

From the one second mark to the one millionth anniversary of the universe there was a swirling dance of nuclei searching for the electrons that would make them whole atoms. When this feverish activity ended all the atoms were complete and the universe consisted of two elements, hydrogen and helium, with hydrogen by far the predominate element.

In the next two stages the ever-expanding and ever-cooling hydrogen/helium gas universe became unevenly dense, somewhat like lumps in a gigantic pot of mashed potatoes. These clumps of condensing hydrogen/helium gas gave rise to a new force in the universe... gravity. This new force pulled the denser areas into large structures called galaxies, leaving behind areas with no gas or voids. Within the galaxies gravity again created new clumps of denser gas. Pressure and heat increased as these clumps condensed until a nuclear fire ignited and they became stars. This colossal change took about 300 million years, yes there were still only two elements in the universe. The situation was ripe for change.

All stars burn with a nuclear fusion fire, a fire about 20 million degrees hotter than a logs burning in a fireplace. The heat and light of this process comes from the fusing of two hydrogen nuclei to form helium. Inside the star the outward pressure of heat and light counteracts the enormous inward pull of gravity on the gaseous mass of the star, and this balancing act of outward and inward forces will go on for 90 percent of a star's life. This was the situation shortly after the birth of the universe, this is the situation now.

Nothing lasts forever. Eventually a star will burn almost all the hydrogen in its stellar body. At that time gravity compresses the core and the temperature rises. With this increased heat a new fusion begins... helium nuclei now fuse into carbon. Simultaneously the outer shell of the star becomes superheated and expands to at least a hundred times its former diameter. Astronomers call this type of star a red giant. Although a star may have burned for billions of years, its life as a red giant will be only a fraction of that time, often a few million years.

Inside the core, the fusing of helium into carbon continues and other heavier elements appear: oxygen, neon, nitrogen, sodium, etc. This fusion process combines to produce the heavier elements, working its way up the ladder of the periodic table. (The periodic table is a scientific classification of all the known elements arranged by their interactive properties and most importantly by their atomic weights.) At this stage the star is like an onion, each fusion fire having created a new layer of elements, the heaviest at the core.

Stellar fusion at the center stops when the star's core has been transformed into iron. While iron is far from the end of the elements on the periodic table, it is the heaviest element burnt in the fusion fires. Yet some heavier elements are also formed in minute quantities at this time.

When a red giant star reaches this stage there is not enough outward heat and pressure being generated at the core to overcome the inward pull exerted by gravity on the outer shells. Collapse of the red giant follows: the outside layers contract inward at fantastic speeds toward the solar center. In this transitional state the heat build up is enormous, fusing the hydrogen remaining in the outer layers. An instant ignition results, blowing the outer stellar layers into space in the form of dust and gas.

The star is now dead. There will be no more fusion at the core; it is now a dense mass of exotic material. These remains are the stuff of scientific theories and science fiction but yield little gold. Only enough gold can be found in this cosmic debris to account for a minute portion of all the gold found so far on Earth. What is needed is another kind of stellar death ...the super nova.

These massive stars, three or more times the size of our sun, burn hydrogen at a faster rate and die with spectacular explosions. These cataclysmic deaths generate more heat and pressure than smaller stars. It is also thought they release great gravitational energy during the collapse. In fact the incredible burst of energy released during the stellar death throe is many times the star's total output during its lifetime. It is in this super-heated 100,000 million degree Fahrenheit explosion that gold and the other very heavy elements are made through atomic fusion.

All the elements on Earth come from the

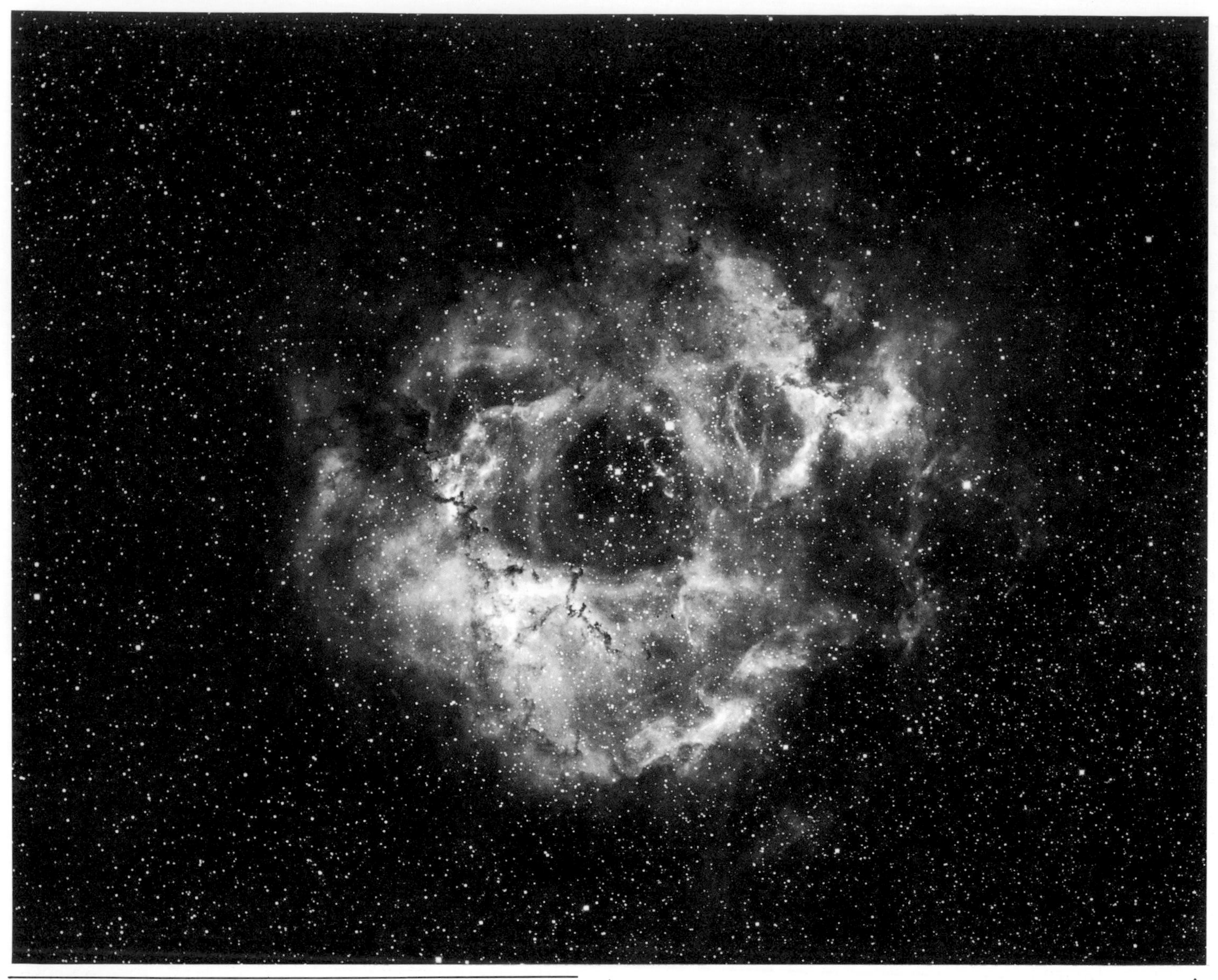

When a star dies in a super nova blast, such as the Rosette Nebula, among the heavier elements created by the extraordinary explosion is gold.

All the elements on Earth come from the death throes of a far away star. In its color and gleam, gold held its moment of birth as a brilliant reflection.

GOLD'S EARTHLY ORIGINS

Our sun is at least a second generation star. It is only 4.5 billion years old, whereas the universe is about 15 billion years old. Some of the cosmic debris that created our sun may have helped to create the Earth.

Some scientists believe the Earth and the other planets in the solar system were plucked out of our sun by the gravitational field of another star wandering by. Other scientists think the planets and sun coalesced from the same batch of cosmic dust and gas. Whichever set of theories is correct, Earth was somehow formed. Like the universe, the fiery liquid planet Earth was hot at its birth. Unlike the universe there were not only two elements on earth but almost a hundred.

As Earth cooled, a rock crust formed. Geologists speculate that the lighter elements were spun to the surface by the centrifugal force of the Earth's rotation, leaving the heavier metals, especially gold, below the crust in a molten state. Further cooling trapped gold in various types of rock formations.

Gases escaped from the cooling rock and created the atmosphere. Later water appeared and then rain. Our planet was bathed in water, which still plays a crucial role in bringing gold to the surface out of the rock formations.

GOLD'S ACTIVE ENVIRONMENT

A living or active planet unlike our sister planets, Venus and Mars, Earth's geology is dynamic, always moving and changing. A large scale example of this activity is seen in continental drift. All of Earth continents rest upon huge granite plates which in turn float above the molten magma deep within the earth. The North American plate is drifting away from Europe and toward Asia. In the middle of Atlantic Ocean there is a huge rift where the continental plates split. Here, deep in the Atlantic, the molten magma is welling up pushing the continents apart. In the Pacific the North American plate dives under the Pacific plate, several hundred miles off the West Coast. It is this action which causes tremors on the West Coast of North America.

At any given moment a myriad of geological changes occur in North America. Water and wind erosion break up mountains. Sediments, sometimes made up of particles from the eroded mountains, settle and form new rock layers. From time to time the land folds over itself, pushing mountains like the Rockies ever higher ...all a by product of the continental drift.

When measured in human life these processes are imperceptibly slow. When measured in the Earth's 4.5 billion year lifetime these changes are like human decades. It is in this geologically active environment that gold rises to the surface of the planet.

GOLD COMES UP FOR AIR

Currently three theories attempt to explain how gold travels from deep under the earth to a shallow level where it is could be readily found. All the theories have one notion in common: the ability of heated liquids to hold minerals in solution. The amount of dissolved minerals in any given liquid is a matter of temperature and pressure.

Many dessert recipes call for dissolving sugar in heated or boiling water. This combination is also a good example of how minerals behave in solution. At or near the freezing point of water, sugar will dissolve up to concentrations of about 64 percent, at body temperature the dissolved amount increases slightly to a 69 percent sugar concentration. But at the boiling point of water the percentage of sugar that can be dissolved jumps to 83 percent. In other words, the hotter the water the more minerals that can be dissolved.

Pressure also influences the concentration of minerals in water. As in a home pressure cooker, pressure will prevent contained water from boiling and turning into water vapor. Water that is still liquid when heated beyond the boiling point and subjected to pressure becomes a superheated liquid that permits heavier concentrations of minerals to remain in solution.

When a superheated solution cools, its ability to hold a high concentration of minerals diminishes dramatically. Minerals and elements then precipitate out of solution in the form of crystals. Gold has a crystalline form and can be carried by such superheated liquids; when the liquid cools, the gold precipitates out of solution.

This first theory goes back to the time of the formation of the Earth's crust. Molten rocks several thousand feet below the surface cool, the less volatile minerals crystallize and form deposits. Gold and the heavier elements are the last to crystallize and are still held in fiery fluid state along with a fair amount of superheated water. With the occasional movement of the Earth's plates cracks form in the crust. Molten material moves toward the surface. The release of pressure sends the less viscous water in fast moving jets. As the pressure and heat decrease silica and many dissolved metals precipitate out of solution and crystallize. Of course the last and most important of these metals is gold.

The second possible explanation for surface gold involves sedimentary rock layers. Sedimentary rocks are the by product of bedrock erosion. Fine grains of eroded material are borne by wind or water and laid down in layers at another location. After many years these layers harden into a new form of rock.

Sometimes the movement of the Earth's plates downfolds sedimentary rocks toward the hotter region beneath the crust. The sedimentary rocks carry a considerable amount of water, as the sedimentary rocks heat up, superheated water escapes and travels toward the surface along fissures, dissolving minerals along the way. Close to the surface the water cools and the gold crystallizes.

The third hypothesis focuses on sedimentary rock forms. But in this case it is the pressure of thousands of tons of sedimentary rock that creates the heat that releases the superheated water and dissolves the gold that eventually crystallizes near the surface. Whichever of the theories is true (perhaps none of them are), gold can be found near the surface of the world.

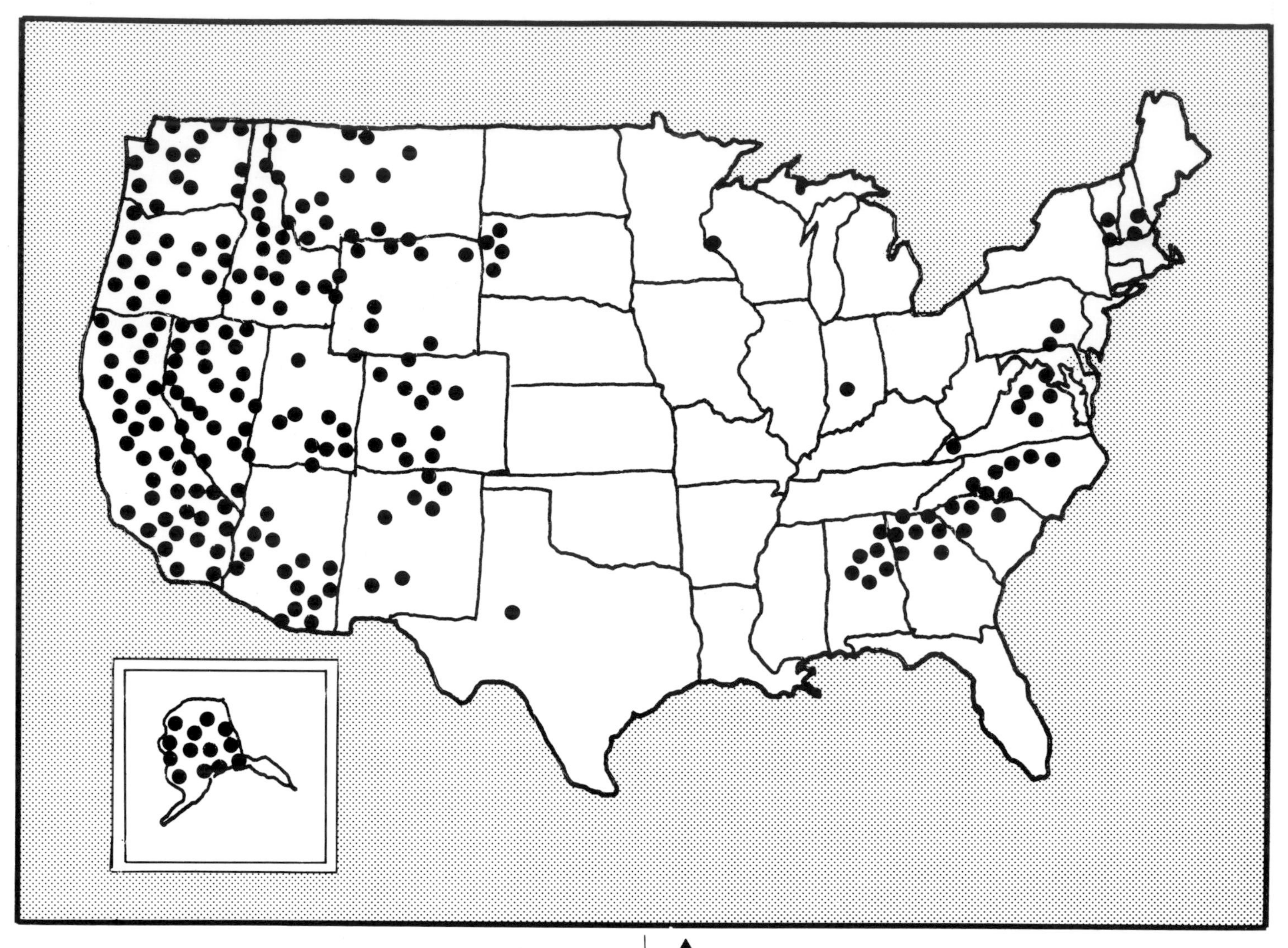

▲
Known gold concentrations in the United States.

CHAPTER 2

GOLD FOR THE PICKING

Some things have not changed since the great gold rushes of the last century. Gold can still be found in the same types of terrain, in the same types of deposits and by the same hit or miss methods. Most of the present day advances in the art of discovering gold have been in the field of geology, the use of metal detectors, or the use of large river dredges. Gold is mined from hard rock formations with the use of heavy machinery made from super-hard high tech metals, not the lowly pick and hammer. But nothing beats hard work and a lot of luck in finding gold.

The two main types of gold deposits are still placer and lode. Lode deposits, as in mother lode, are veins of gold that must be tunneled or mined. Placer deposits are mostly found in streams and are the eroded remnants of lode gold. Placers can be recognized easily by their smooth, worn appearance. Even though lode deposits will yield more gold in the course of mining, placer deposits are the more important when discussing the North American gold rushes.

LODE GOLD

The fabulous gold mines of history and myth were all lode gold deposits. In a sense lode deposits were reserved for those few gold rushers who had real mining skills and knew exactly what they were doing. Finding the mother lode demanded time and patience. Digging shafts and tunnels that would not collapse demanded expertise. Getting at the gold was just the beginning in a lode discovery.

Buried veins of gold streaking through other materials, usually granite or quartz, are hard to mine and retrieve from the mined ore. There is so little gold compared to the rock that holds it. Yet much of the gold can be extracted through precise processing techniques, with the right machinery and skilled labor.

PLACER GOLD

Retrieving placer gold is much like entering a lottery. You don't have to know much to enter and the odds are only marginally better. Placer gold was, and still is, the glittering prize at the end of our get rich quick rainbow.

In live stream beds placer gold is hidden very near the surface and there for the picking. The lucky key to finding placer gold is one of gold's major attributes, density.

Through weathering, earthquake tremors, floods and the like, the material holding the lode gold is eroded and released along with the gold on the mountain or hillside. These eroded rocks, minerals and gold are washed into streams, rivers or occasionally out to sea by the rain. Gold, being heavier than the other eroded debris, travels a shorter distance and sinks at the first available opportunity to the lowest possible point.

As a stream or creek flows it deposits gold nuggets, flakes or flour in places where the current is interrupted or slowed. A subtle area to find gold is at any point where a stream abruptly widens. It is here that the water flow slows downstream and the velocity is not fast enough to carry heavy materials like gold. Sometimes a small sand or mud bar will help mark the spot.

A more obvious placer bed is the gravel area around the bend of a fast moving stream. Here the water flow is also decreased by the obstructing curve in the creek, but it is now deflected to the opposite shore where the water velocity is then increased. All around the bend slower moving water descends toward the gravel and pebbles in the curve. This material traps other heavy material like gold, but lighter debris is washed away by the slow moving current.

The most conspicuous hindrances to downstream flow are logs and boulders. In this case the large object forms a breakwater that creates a pool of slower moving water on the downstream side. Some of the gold carried by the water is deposited in this quiet area, protected from the strong current. An outcropping of rock will act in lesser manner as an obstruction and gold can frequently be found on its downstream side as well.

Ancient streams are dried up streambeds. One can be fairly certain to find these relics on the dry banks immediately above a present stream. Nothing erodes the land like moving water; given enough time, it cuts channels as deep as the Grand Canyon. But the cutting action of the water follows the contours of the softest underlying material. Layers of clay or sedimentary rock will be cut away by water erosion before a layer of bedrock. But again, given enough time, water will cut a channel even through granite.

Often a stream was once a much wider river where the water eroded a v-shaped channel through the rock to its present creek bed. It is high on these walls of the older and wider channel that placer gold deposits can be found. When placer gold is found mixed in gravel from the ancient stream it is called a bench placer.

Deep canyons with walls 20 feet high or more show different layers of history. Often there will be a striation of gravel on one of the walls. This gravel patch may have been a former bend in the river or stream and some placer gold may have collected at that spot eons ago.

Some streams may have had their course changed more cataclysmically by rock slides, earthquakes or even volcanos. Here the ancient stream is usually covered with weathered stones and debris; in some cases the bed has been covered with lava. Yet under all the fill material lie tiny pockets of gold where bends, wide berths and obstructions once channeled the flow. These ancient stream placer deposits are obviously the hardest to find and require an almost monumental job of excavation. But the prospector who finds a buried ancient stream can be sure no one else has scoured his site before.

▲ *Possible gold locations. 1.–Moss that grows along the creek and on stationary objects. 2.–Gold could hang inside bends of the creek where sand and gravel bars have built up during high water. 3.–Behind or on the downstream side of boulders, in pockets or potholes formed by the stream's current. 4.–Natural breakwaters. 5.–Just above any natural shelf. 6.–Tree roots can trap gold. 7.–Grass roots along water's edge. 8.–Cracks in bedrock of ancient benches that were beds of early waterways.*

Found mostly in mountain streams, crevice placers best illustrate gold's affinity for the lowest point.

Bedrock is never uniformly smooth. It often contains cracks due to the earth's movement and erosion. As gold bearing water runs over these crevices, a small percentage of the gold drops into the cracks. However, the crevice placer is also filled with stones and pebbles and whatever else the rushing water has wedged into the crack. But again gold, being so dense, will work its way to the bottom of the crevice, helped by the agitation of the flowing water. Here the gold waits for some lucky and industrious prospector to pry loose the covering stones and bring it to the light of day.

It is neither lode nor placer gold but a bit of both. Pocket gold occurs only in combination with quartz, mostly veins that once ran perpendicular to the horizon. As with placer gold, the holding quartz vein disintegrates over time because of re-

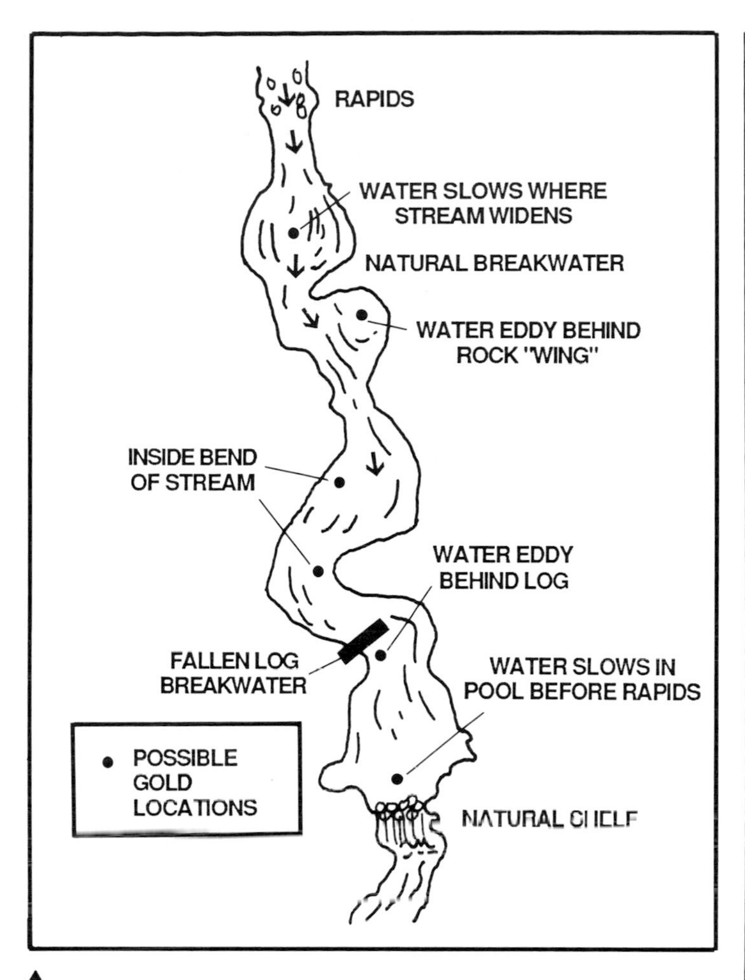

▲
Where Nature hides her gold.

peated tremors, ground shifts, earthquakes and other vibrations. The fragments of quartz and gold are washed away by the rain. Once the quartz is gone, the remaining gold falls into the tunnel left by the decomposed vein. The gold that falls into the hole has never been smoothed by tumbling or rolling in a stream; it is sharp and jagged. This is a sure way to tell pocket gold from its waterwashed brethren

Don't think that once quartz has crumbled you are left with just an empty hole. The disintegration and washing away process occurs at the same time as earth, dust, small leaves and twigs are carried into the hole by the rain. One indication of a decomposed quartz vein is a soft area in normally hard ground.

When gold bearing quartz veins of over 50 feet disintegrate, more than one pocket will form. Each pocket but the last will form at the narrowest part of the decomposed vein, much like a series of hourglasses end-on-end, creating a long tunnel. The last pocket, of course, will be at the bottom of the once quartz-filled tunnel.

The relationship between placer gold (the eroded remnants of a lode deposit and lode gold) and the hard rock veins of gold and quartz, are very important to any gold seeker. The presence of placer gold gives a strong indication of the existence if not the whereabouts of the mother lode. In 19th century North America, many fortune hunters were content to work the stream beds and gather, if not enormous quantities of gold, enough to make their ordeal highly profitable. Others were hell bent to make the really big strike. The mining of gold is central to an understanding of each of the gold rushes.

MINING THE GOLD

Will it pan out? Does it pan out? See if the idea pans out. Panning, the single most important concept in gold prospecting, has coined a colorful idiom in American English. Some form of panning was in use about four thousand years ago; the principles have not changed since then, only the pan materials.

Density. Once more we return to gold's incredible density. Gold is heavier than virtually all the other materials it is associated with. Almost all the recovery processes for placer gold are based on the idea of density combined with the recurring phenomena of running water and agitation. The first step is always to find where the placer gold is hidden; and the simplest and easiest tool for the job is the pan.

PANNING GOLD

Any plain, round, flat-bottomed pan, 12 to 16 inches in diameter at the top, tapering to between 5 and 9 inches at the bottom, with 2 1/2 to 5 inch sides will do. It is said that many a spur of the moment gold rusher used his frying pans to great advantage.

Traditionally, steel pans were used. Modern pans are made from a thick green plastic, which allows gold to standout visually on the bottom, does not attract mercury, and is non-magnetic. Yet "old timers" eschew the plastic pan claiming it is less rugged. You need a durable pan when you are traveling on foot in a mountain range, many miles from the nearest road.

The steel pans of the last century had to be fire cured frequently and scoured to rid the inside surface of grease and oil. Oil floats naturally in water and will attach itself to small flakes of gold. Despite gold's great weight, oil and grease float small golden particles to the top of a pan submersed in a rushing stream and many tiny pieces of gold are lost over the side. Scouring the inside of the pan will also create a rough surface that catches small flakes on the textured bottom. Now, equipped with the perfect pan, our gold rusher must seek a promising location, preferably

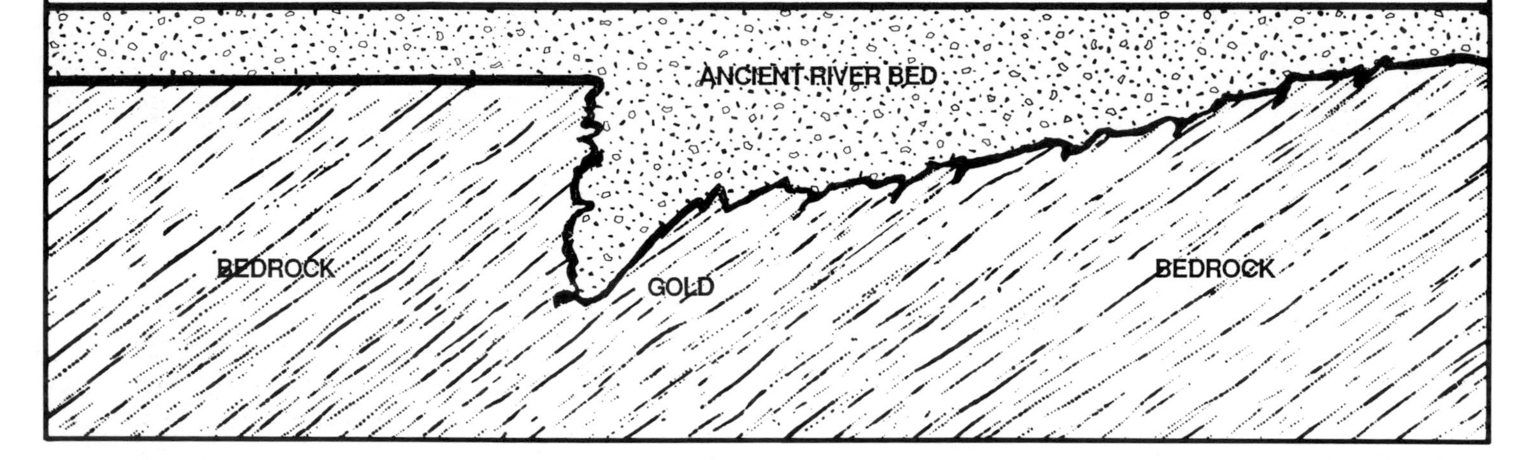

▲

Cross section of a river bed showing down faulting on the upstream side, permitting black sands, gold and silt to accumulate in this natural trap.

▶

Where the nuggets come from–residual placer from decomposed quartz lode.

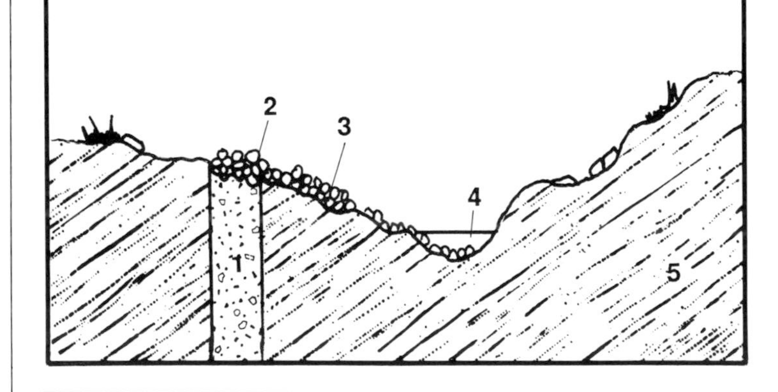

near a source of running water.

You pan gold to extract it from the extraneous materials that fell into company with the gold. We are talking about finding placer gold and, you will remember, most often the place to look is in a flowing steam or small river where the bed widens or bends. The gold may be mixed with gravel, it may be mixed with clay, or flakes of gold may be coated with silt. Wherever the gold lurks, panning will expose it.

Having found a suitable body of running water, the next step is to take a soil sample by digging as close to the bedrock or hard clay bottom as possible. This probably means pulling a sample from the bottom of the gravel pile. With our sample in the pan, we begin the panning process.

Much has been written about the correct method to pan for gold. Every aficionado has one. But all the techniques have one principle in common: the agitation of a pan full of water so the gold will separate from all other material and fall to the bottom where it can be recovered.

Over a long period of time, panning can be arduous. First the prospector settles in an area where the current is very slow. On a summer's day the cold water may be refreshing; later on it will numb and cramp the leg muscles. The miner covers the sample in the pan with water. Cautiously all the clay or earth lumps are broken up. All root matter is discarded after being carefully washed to retain any gold flour in the pan.

The remaining mud, gravel and pebbles are thoroughly mixed with the water. This mixture should resemble a mud soup, not a mud pie. Slowly, so as not to disturb the contents of the pan, the pan is submerged about 1 or 2 inches below the surface of the stream. Now, any technique which evenly rocks the pan is applied to the mixture of wet, prospective gold bearing mush. Old timers do this with either a side to side motion, a back and forth motion or a swirling motion.

All the lighter debris will lift out of the pan and float into the slow moving current never to be seen again...which is a good reason for doing this

▲
The process that was used in ancient times to separate the wheat from the chaff was exploited in waterless areas to separate the gold from the rocky soil.

step very carefully. No gold hunter wants to think that the main purpose for squatting in a chilly creek all day long has just floated downstream.

For a brief moment the pan is lifted out of the water and all the large pebbles are individually inspected to see if there are gold nuggets. All that does not glitter is returned to the stream and all that does glitter is pocketed. The pan is re-submerged.

The process begins anew and is repeated until all the larger pieces of gold are retrieved, or most of the large extraneous material is washed away. What remains after this stage is several teaspoons of black sandy matter.

CONCENTRATES: TAILING UP, ROASTING AND POTATOES

At this point most people just stare at the black sand at the bottom of pan and wonder if all the work was worth the trouble. Chances are the search has been fruitless. Yet it would be a grievous error to stop here. The gold hunter's true skill is in reading the black sand at the bottom of the pan and identifying "the concentrates."

The first indication of gold in the sample is tiny flecks of the bright shiny particles in the concentrate. These gold particles are sometimes called "colors," and the process of exposing the colors is called "tailing up."

Once more gold's density leads to its recovery. Tailing up is similar to panning with one major difference: the pan of concentrates is not rocked

▲ *Rockers, essentially, were simple to run. Although easily operated by one person, it was more profitable to divide the work among two or three men.*

in any manner; it is briefly swirled. A minute amount of water is added to the pan and gently swirled to form a comet action that will move the lighter particles ahead of the heavier material. When done correctly the concentrate sorts itself by weight, grading from the lightest to the heaviest. If the sample contains gold, it would show "colors" at the tail of the swirl...hence the expression "tailing up."

Several other methods are used to further distill the concentrate. An obvious but time consuming approach is to pick out gold particles and flakes with tweezers. Or when iron-rich particles tail out with the gold, you minimize the iron based debris by running a magnet over the dried tailings. Some regions are rich in sulfides that tail out with gold or lock up fine particles of gold in larger sulfide pieces. These make it next to impossible to determine if the sample just panned contains any gold at all. However, there is an old gold rushers trick to solve the problem: "roasting."

Roasting the concentrate does what the name implies. It cooks the sulfides out of the debris. The concentrate is dried and then transferred to a metal flat surface. In the last century, more often than not this was a coffee tin top. The lid is placed on a hot fire and allowed to burn until no more sulphur and arsenic fumes are given off. Any gold hidden in sulfide tailings is now freed.

When very fine flour gold is picked up in the tailings, another old-time trick of the trade is used, the mercury potato. Mercury amalgamates with fine flecks of gold only if the gold is free of mineral coatings and grease or oil, and these mineral coatings can often be burned off in the roasting process. Yet some oils such as pine oils

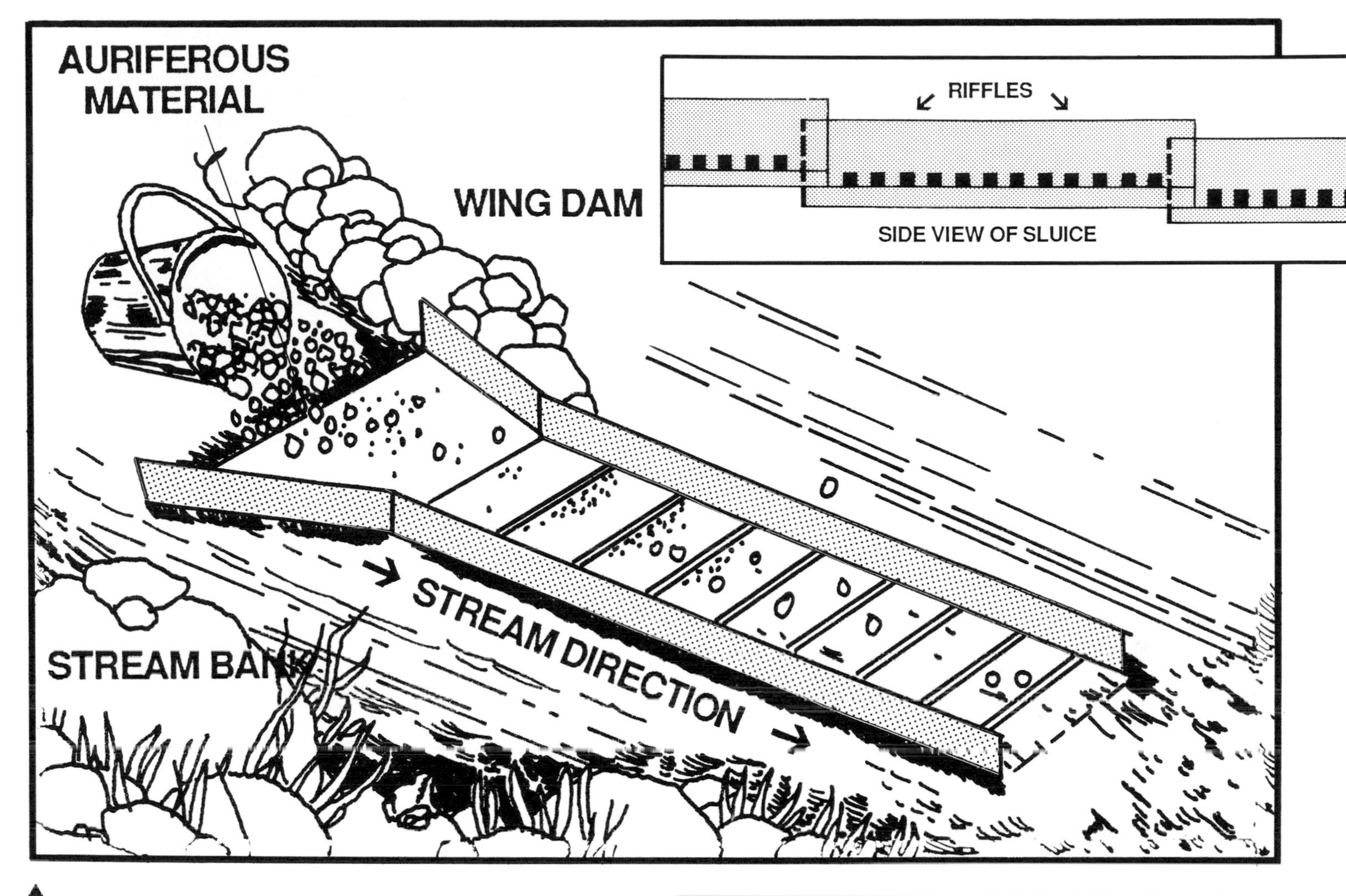

▲
Hopper end of sluice box is fed auriferous material for classification.

are found naturally and others are transferred to the gold from oily or greasy pans, which is the primary reason the pan used in panning must be scoured and burned to remove any oil and grease. When large amounts of concentrates are gathered by using sluices and rockers, lye is added to the holding buckets to dissolve minerals and oils. To pick out the gold flour, a teaspoon of mercury is added to the concentrate and stirred with a small iron rod. Old timers believe the iron rod helps mercury to join with gold. When this task is done a silverish-gold pellet will stand out from the black sand concentrate.

Now our baked potato is used as a retort to distill the mercury from the amalgam and leave behind a small valuable button of gold. The procedure is simple. First the potato is cut in half and a scoop, about the size of the amalgam, is taken out of one side. The two halves are wired together and put on top of a small fire for an hour or so. Great care must be taken not to inhale the mercury fumes, as they are poisonous. Next, the potato is removed from the direct fire and buried in ashes that are hot, but not glowing, for another hour.

When the potato skin is blackened and a stick easily penetrates the meal, the potato is removed and left to cool. To retrieve the gold all that remains is to open the potato and remove the nugget. Some prospectors also squeeze the potato to retrieve their mercury. The now poisonous baked potato must be buried so no person or animal accidentally eats it. Eating this potato will result in an unpleasant death.

LARGE SCALE PLACER MINING

All panning really does is locate a gold rich area. One of the immutable laws of gold prospecting states: the amount of gold recovered is directly proportional to the volume of material processed.

An intelligent miner figured out that if a wide spaced screen was placed inside the pan above the bottom, most of the finer material could be easily filtered out of the large gravel. This utensil is a gold pan sieve. In operation the gravel is shoveled onto the sieve and a pail of water is poured over the sample. The finer material falls into the pan, leaving the gravel to be picked over

ALBUQUERQUE ACADEMY
LIBRARY

in case there are any large gold nuggets hidden within.

Another important device is a riffle. It comes in many shapes and sizes, its designs are limited by the imagination and it is used on all large scale gold retrieval machines. A riffle is a series of grooves, channels, slats, or wire screens in a sluice box that catches gold.

On a larger and more primitive scale than the sieve is a crude device aptly named the log washer. It is nothing more than a hollowed out log with a 3 to 5 foot handle attached lengthwise. At the bottom of the log is placed a riffle. The miner puts several shovels of gravel in the log washer, followed by several buckets of water. Using the handles the miner rocks the log to create a dasher motion. The water and concentrate wash out the riffle end and the gravel eventually works its way out the other end. The log washer, although easy to make and operate, risks losing valuable gold along with the gravel. During the Georgia gold rush in the 1830s someone grew tried of using the log washer or suspected he was losing too much gold and invented the legendary rocker.

A typical rocker consists of a washing box with a screen, a cloth apron under the screen, and a small sluice with several riffles and rockers attached underneath. The bottom of the washbox is made from either a plank with 1/2 inch holes drilled into it or a wire mesh with 1/2 inch openings. One or two shovelfuls of wet gravel are placed in the washbox. Water is poured over the gravel and the rocker is gently swayed. After the gravel is washed clean, it is inspected for gold nuggets and the large debris is removed. The smaller substances along with any gold are caught in the apron. The riffles are there to stop any gold that might spill over the apron. Although one man could work a rocker, it was usual to find two or three working in tandem. One man would dig the gravel and bring it to the rocker; one man would get the water and wet the gravel; one man would do the actual rocking. This way a lot more material could be processed.

Another innovation also increased the amount of material processed and decreased the workload. But the Long Tom, as this device was called, needed a steady supply of water. It is basically a two section inclined trough. The first section is about 10 to 12 feet long and widens at the bottom as it meets the second section. The second section consists of a perforated iron sheet that acts as a sieve, plus a 5 to 6 foot long riffle box under the iron sheet to catch the concentrate.

When water is plentiful, sluice boxes are the answer to large scale processing. Sluice boxes are a series of inclined trough sections with each section containing a series of heavy ribbed riffles set perpendicular to the flowing water. Gold bearing debris is dumped into the upper end of the sluice and flowing water carries the material down the length of the sluice boxes. The lighter matter is washed out the bottom. Heavier material and concentrates are trapped by riffles. When the riffles fill up, the water flow is stopped and the riffles are cleaned out.

A miner with a pan could process about 1/2 a cubic yard of gravel a day. The log washer could process almost 1 1/2 cubic yards of debris a day. With a rocker two miners could process about 4 cubic yards of material a day. The Log Tom could double the output of the rocker. But the sluice boxes, when there was enough flowing water, could process 20 to 40 cubic yards a day. The location, the experience of the miner, and whether or not the prospector was willing to go into partnership with other miners determined which device to use.

As both the Long Tom and the sluice boxes were labor intensive equipment, it really came down to the age old question: "Whom do you trust?"

DRY WASHING

Not all gold is hidden near a current source of water. Arid areas in North America once had running streams. These streams carried eroded lode gold and deposited it in much the same manner as wet stream placer gold. The problem is not one of finding the gold, but more of processing the samples. The technique for processing gold with little or no water is called dry washing or dry panning.

The dry processing of gold depends again on the main characteristic of the metal, its density, and substitutes wind for water. Winnowing is one of the oldest processing concepts and dates to man's prehistory. Instead of separating chaff from grain, miners winnow gold from the surrounding fine rock particles. All that is needed is a heavy wool blanket and a strong breeze. First, the gold bearing material must be very dry; second, after all the large gravel is removed, the ore must be pulverized. Next, the dry concentrate is put into a blanket and carefully tossed into the breeze so the lighter matter is blown away. Because of gravity the heavy gold will fall to the blanket where it will be caught in the weave. Later, the gold dust can be shaken or picked from the blanket.

There is also a dry washer for use in arid locations that resembles a sluice box complete with riffles. Instead of using water, the dry washer is vibrated so the lighter material falls out one end and the concentrate is trapped by riffles.

ALL THAT FOOLISHLY GLITTERS

Fool's gold is any material that resembles gold enough to fool the neophyte. These masquerading minerals are usually iron sulfides such as pyrite, pyrrhotite, chalcopyrite and marcasite. A sure test is to strike the suspect material with a hammer. If it is not gold it will shatter. The most malleable and ductile of all metals, gold is not brittle and will stretch before it breaks. Gold never shatters. When found in a relatively pure state pyrrhotite, chalcopyrite and marcasite can be picked out of a sample of concentrate with a magnet. Gold is not magnetic. All of the iron sulfides will tarnish when exposed to air and/or water. Gold does not corrode or tarnish when exposed to oxygen, water, salt or any other naturally occurring materials. If a gold hunter is fooled by the iron sulfide he has not done his homework.

Mica can sometimes be mistaken for gold. Generally it is out of frustration that a novice prospector will make this error. Mica's density is so much less than gold that most of the time it is washed away before it can be noticed. Only after hours of fruitless endeavor will a prospector begin to mistake flecks of mica for his nascent fortune.

Techniques have changed very little since the gold rush days. Today many weekend prospectors will go into the mountains with plastic pans and follow much the same procedure as the gold hunters did a century before. It was, and still is, arduous work, sometimes for little reward. Gravity and density are the keys to gold processing. All the techniques exploit these two fundamental givens. Because such small pieces weigh so much, a small find is a huge reward.

▲
River mining. Here, miners are building a diversionary channel. The river water is let in by means of a closeable gate. With controlled flow, the miners were able to create, in effect, a large-scale sluice.

FRANK LESLIE'S
ILLUSTRATED
NEWSPAPER

o. 1,236—Vol. XLVIII.] NEW YORK, JUNE 7, 1879. [Price, with Supplement, 10 Cents.

Ever since 1849 newcomers to the gold diggings were considered to be totally unprepared for the hard life of a placer miner. In this 1879 exaggeration the "swell" reflects the attitude of a laborer and investor rather than that of the Argonaut and experienced miner.

CHAPTER 3

A GOLDEN GLEAM IN THEIR EYES

From the very beginning the history of the Americas has been woven with golden threads. No one is really sure about the motives of the pre-Columbian European and perhaps African visitors, but one of Columbus' goals was gold.

As soon as Columbus realized he had discovered a new land he started asking the natives he found there about gold. His preoccupation with gold appears throughout the journals of his voyages. When he brought back to Spain word of his discoveries the rush was on. Foremost among the workers arriving in the New World were skilled miners.

THE SPANISH RUSH TO THE AMERICAS

Sensing there was great wealth to be gained in the Americas, Spain obtained a series of papal bulls from the conveniently Spanish pope, Alexander IV, and as a result Spain and their rival in the New World, Portugal, signed a treaty dividing up the Americas and the East between them. For almost a century the rest of the world outside of Europe seemed to belong to the Spanish and the Portuguese. Spain had the Americas, with the sole exception of Brazil, and the Portuguese had all of Asia, the East Indies and Brazil. The rest of Europe ignored the pope's decision. They started to colonize the Americas less than a century later to challenge the Portuguese holdings in Asia. The first were the Spanish, and high on their agenda was gold and silver.

In 1499 the island of Hispaniola, which comprises the modern countries of Haiti and the Dominican Republic, was the first site to be mined by Europeans in the Americas. The supply of gold on the island lasted less than 20 years. During its most prosperous year 60,000 ounces of gold were mined. In 1513 gold was found on Cuba and many of the miners along with the governor on Hispaniola rushed to the smaller island to seek their fortunes. Henceforth Cuba became the Spanish stronghold in the Americas. These Cuban mining endeavors also lasted about 20 years, with much the same success as on Hispaniola. Miners, encouraged by the Spanish Crown, spread throughout the Americas in search of gold. But ahead of them were a different kind of man.

Leading the search for gold were the adventurers. They had names such as Cortes, Pizarro and Quesada and conquered Indian civilizations in Mexico, Peru and Colombia. The Spanish called them conquistadors. These were men trained in the new Spanish infantry discipline and almost unbeatable in the field for over 100 years. In the 1500s their only European military rival were the Dutch whose expertise lay in employing citizen soldiers behind ingenious fortifications. It was the only countermeasure against the Spanish infantry that worked at the time. Against such men who had excellent leadership, a strong sense of discipline and incredible endurance for privation, the Indian civilizations of the Americas were in a losing position, and lose they did. The Indians lost everything, not only their gold, but their freedom.

They came for gold and their method was simple, plunder. At first the going was tough for the conquistadors. In 1509 Alonso de Ojeda led an expedition into northern Columbia and was roundly defeated by the natives. Another expedition went to Central America and was routed by the Indians. But the tough kept going and two years later survivors of Ojeda's forces established a small outpost on the Isthmus of Panama. Under Balboa this quickly grew into the small city of Darien. A short time later, in 1519, Hernan Cortes set sail from Cuba with 11 ships, 550 men and 16 horses. His objective was to find the rumored Indian gold hidden somewhere in Mexico. Within three years of arriving in Mexico he toppled the Aztec empire, gained a fortune in gold and silver, became governor of New Spain and a legend in his own time. A few years later the lure of gold lead another man, Francisco Pizarro, who was also a survivor of Ojeda's ill fated expedition, to topple the Inca civilization and enrich himself with gold, silver and jewels in the process. Other Spaniards would follow suit, looking for luckless Indians to plunder. None would reach the heights of fame gold and bloodthirstiness that Cortes and Pizarro reached.

The search for gold lead the conquistadors into the jungles, mountains, deserts and plains of the Americas. Following right behind these military supermen were the miners who ventured even deeper into the unknown land. Where the conquistadors had the advantage of superior military training the miners had the advantage of royal protection. The Crown did everything in its power to encourage mining. To this end all miner's possessions were exempt from any form of legal foreclosure. They received special treatment by law enforcement officers and were accorded special leniency in the courts. By 1550 the importance of plundering civilizations and robbing graves ceased to be important. It became time for the miners and colonists to take center stage.

Spanish mining methods were used extensively during 16th and 17th centuries in the Americas. Basically this meant the Indians were enslaved and forced to work the mines while the Spanish miners who had the expertise became the supervisors. At first the miners searched for only the richest ores and used an age old simple and wasteful smelting process to extract the gold. But in 1556 an amazing discovery was made at the Real del Monte mines in Mexico that would revolutionize the gold mining industry. Some unknown miner figured out how to use mercury's ability to amalgamate with gold in gold mining. This quickly lead to using mercury to recover gold from low grade ores; which lead to a dramatic increase in gold production. All this greatly pleased the Spanish Crown because they owned the chief world source of mercury in the mines of Almaden, Spain.

▲ *In the early 1850s it seemed as if representatives from the four corners of the globe had come to seek their fortunes in California. Notice the Native Americans in the left-hand portion of the picture. They are not participating in the gold rush, but rather leaving their homelands and venturing on a road less traveled.*

The exploration of much of the Americas was accomplished by the conquistadors. They were pulled by gold's hypnotic attraction, a force which acts on human desire and turns it into greed. Even if they did not find gold in their exportation of uncharted lands, they blazed a trail for the miners in Mexico, Central and South America. Wherever the miners dug the colonists went, and wherever the colonists went they built cities, and wherever they built cities farms and industries grew, and wherever farms and industries grew trade increased, and wherever all this activity occurred the land was settled.

When all is said and done, the conquistadors plundered their way into the history books and the miners, who found the enormous quantities of gold and silver in the ground the conquistadors walked upon, are often overlooked. Although there are no dependable records of how much gold was mined in the Americas, the best estimate is about 4 billion in 1985 dollars. All the conquerors' combined plunder barely totaled 400 million in 1985 dollars. As in the later gold rushes in North America the real winner was the people and the land.

Gold, the strange and valuable element born of the fusion process at the spectacular end of a star's life, is a force in human history that brings about the fusion of society. After the miners came and worked the earth, the farmers, the tradesmen and the ranchers came to Spanish America. By 1800 the land and its products had become the leading economic force in Latin and South America, outproducing, in terms of profit, all mining endeavors by 4 to 1.

If there had been no gold, no rumors, no stories

told of legendary golden sunlit cities that gleamed with a metallic luster, the extensive Spanish exploration of the Americas would not have occurred. There would not have been the driving force of golden greed that draws men into hot steamy jungles or bone dry deserts or cold inhospitable mountains. The Spanish would have been content to spend their days by the coastal towns and river cities. Yet there was gold to be found, and it proved to be a mixed blessing to Spain.

Supply side economists look at 16th century Spain as proof of their theories. In the beginning the influx of valuable metals seemed to be a boom to Spain's economy. As gold and silver poured into Spain the supply of these precious metals grew at a faster rate than the volume of goods and services. This caused a rise in the demand for goods and services, which made the prices rise and an inflationary spiral begin. Some of this inflationary pressure could have been forestalled if the Spanish had invested their new found wealth in the production of goods and services. But this was not thinking of Spain in the 16th and 17th centuries. Most of the new shiny metallic wealth from the New World was spent on goods, services and political objectives rather than investment in new industries.

For example, a Spanish noble has an income of $100 a month and his butcher charges him $10 a month for meat. The butcher buys his meat from a rancher for $5 a month. These prices of course are very hypothetical. Now the noble has $1,000 a month income due to a fantastic increase in his wealth through the influx of gold and silver. The butcher sees the noble's new wealth as a chance to make some more money for himself and he raises his price to $20 a month for meat. The rancher also is aware of all this new money that the noble and the butcher are enjoying and he raises his price to $10 a month for his produce. Although the butcher is making more money he is now paying more for his meat, so he raises his prices again in order to earn the amount of money he felt would make him comfortable. The noble, who is still receiving a share of the fabulous gold and silver revenues, balks a little but pays the price increase.

Under these circumstances the supply of available goods and services remains the same, but the prices continue their upward spiral. Had the noble invested some of his wealth in a form of production, such as a new ranch, for example, he would have created a new supply of meat and a competitive force for balanced pricing. The upward inflationary spiral could have been averted with capital investment. That did not happen, and by the beginning of the 19th century Spain found itself with very little money in the treasury and losing colonies to revolution in the Americas.

CHAPTER 4

THE EARLY GOLD RUSHES

The Spanish rush to the Americas does not fit the classic image of a gold rush. Because Spain was interested in developing the wealth for the use of the Crown, only Spaniards were allowed to partake in the profits from mining the gleaming prize. Theoretically, a tax of 20 percent of all gold and silver went to the Spanish Crown since they owned the mines in the first place. This was called the "royal fifth." No one knows how much unaccounted for gold was smuggled out of the mines under the eyes of the royal tax collectors. The Crown suspected this and frequently made mining contracts that exceeded the royal fifth to compensate for the expected smuggling.

Europeans were not welcome to participate in the gold rush, yet some European sea going powers found a way around this prohibition. They turned to piracy. Throughout the 16th and 17th centuries the scuttling of Spanish ships, the murdering of Spanish seamen, the ravaging of Spanish ports and the seizing of Spanish gold and silver was considered to be an honorable profession by the kings of England, Holland and France. Perhaps it was the exclusive nature of Spain's rush to gold in the Americas that brought about this reaction from the other European monarchies. The first real gold rush in the Americas took place at the end of the 17th century and it would set the stage for others to follow.

THE BRAZILIAN GOLD RUSH

The Portuguese bumped into the New World while looking for a sea route to India. In 1500 the Portuguese captain, Pedro Alvares Cabral, already knew there was a large land mass somewhere to the west of his 13 ships. But Cabral claimed to have been blown off course by unexpectedly strong winds, and the next thing he knew there was this strange green jungle covered land with naked Indians running about. Cabral's men discovered a strange red hardwood similar to the dyewood imported from the Far East, long known in Europe as "brazil." The name was applied to wood found in this new American land and a short while later it became the name of the chief Portuguese in the New World.

At first the Portuguese were indifferent to their large acquisition. The Court of Lisbon was more interested in the profits from trade in India's silk, spices, jewels and collecting tribute from Far East cities than this jungle matted territory thrust upon them by accidental discovery and papal decree. This South American land did not have cities and grand palaces. To the Portuguese it was a very uncivilized place where they suspected the natives practiced cannibalism. However, events were about to occur that would change the Court of Lisbon's mind about Brazil.

The fabulous Portuguese trade with India and

the Far East became increasingly unprofitable. A great part of this trade was based on spices and the European market was glutted with pepper, cloves and cinnamon. From a consumer's point of view there is a definite limit on how much spice anyone can consume. At the same time, trade in the East Coast a fair amount of Portuguese lives. Although partially fortified trading settlements were set up in Bengal and China with the consent of native rulers, in India and Ceylon the trading posts were really forts and quite often suffered attacks from the indigenous population. There were also limits on how much Portuguese blood could be spilled for an avaricious Court at home.

Spain's dazzling foothold in the Americas coupled with Portugal's cautious, if not rightfully paranoid attitude toward her ambitious Iberian neighbor, made a counterbalancing colony in the New World a strategic necessity. In line with the same global strategy was the French interest in the under-colonized Brazilian lands. King Francois I of France found the Iberian kingdom's claims to a monopoly in the New World, which were Papally endorsed and arranged, to be an incredible and unacceptable turn of events. He said, "I had never seen a clause in the last will and testament of Adam conceding such control to the kings of Portugal and Spain." To this end he began a series of raids on Portuguese shipping to and from Brazil, and at the same time sent his own missionaries into the jungle land. All this did was to firm Portugal's resolve to hold on to and develop Brazil.

The Portuguese had a large colony in the Americas and Brazil began paying off in unexpected dividends, dyewoods. Europe's growing textile industries were hungry for dyes and the red hardwood discovered by Cabral's men helped to fill that need. Sugar was introduced in 1532 and its cultivation grew by leaps and bounds until competition from the British and the ineptitude of plantation owners brought about a decline in the crop's profitability.

Agriculture turned out not to be the Portuguese strong suit in the colonization game. They lacked the farming skills and common sense. Languid growers destroyed forests which kept the soil fertile, knew next to nothing about fertilization and choosing seeds and neglected to combine raising stock with crop cultivation. In the atmosphere of this dismal agricultural failure the colonial rulers prayed for a golden miracle. As early as 1542 one of the Portuguese governors wrote the king: "...As for gold, My lord I never cease to inquire but it would be necessary to go through sections of very perverse and bestial people to reach lands where gold is supposed to exist." Their prayers were answered 150 years later.

To discover gold Brazil needed men who were willing to venture into the inhospitable jungle, who could brave and withstand Indian attacks and live by their wits alone. What was needed was men who could disappear into the interior of the green hell for years at time and think of it in much the same way we would consider hiking in the woods for an hour or two. What Brazil needed to discover gold in the jungle was not the colonial Portuguese but a native Brazilian and a particular kind of Brazilian at that.

The Paulistas were a very peculiar breed of Brazilian. They were born in the town of Sao Paulo, the offspring of Portuguese fathers and Indian mothers. The Paulistas were raised with knowledge of three cultures and two languages. From their fathers they learned colonial culture and language of the Portuguese, from their mothers they learned the Indian culture and language of interior natives and from their environment they learned the outlaw culture of Sao Paulo. The region and city of Sao Paulo was isolated by a large mountain range from the mainstream Portuguese colonial life. It was an excellent haven for criminals and they made use of often when the colonial government was looking to throw them into jail. None of the colonial soldiers and their commanders wanted to venture to Sao Paulo unless it was absolutely necessary.

Time has a way of romanticizing the truth. In the later half of the 20th century popular Brazilian history has painted an idealized picture of the Paulistas. They are wearing heavy coats, dark wool trousers, high jack boots and wide brimmed "pilgrim" hats. Only the colonial officials and visiting Portuguese royalty would follow the dictates of Court fashion in a tropical climate over the obvious functional dress of the natives. The real Paulistas went barefoot into the jungle. They wore simple cotton loose trousers with an occasional cotton shirt, beard and broad-brimmed hat. In other words they wore the bare necessities and sometimes what they wore was not always picture book clean.

Before they discovered gold in the Brazilian interior the Paulistas earned their livelihood tracking down slaves. It did not matter whether these slaves were the native Indians, perhaps distant relatives of the Paulistas' mothers, or runaway black slaves, recently kidnapped from Africa. To this end they went into the interior armed to the teeth. Typical weaponry for each Paulista in a band of slave trackers, also called *bandeirante*, was a sword, two pistols, a knife, a rifle, cartridge belts loosely hung about the body and a bow and arrow. They were reputed to be better marksmen with a bow and arrow than the Indians. When the Paulistas went into the jungle they meant business, and the slave business was good.

Some *bandeirantes* consisted of as little 10 men, some had as many as 1,000. Wherever Paulista went they established settlements at watering holes, river stops and later at gold lodes and diamond discovery sites. The Paulistas traveled on foot and by canoe, but never on horseback. Horses were an incumbrance in the wild and inhospitable interior. Slaves and canoes carried supplies with greater efficiency. Natural boundaries of rivers, deserts, jungles, mountains and swamps were challenges to the Paulistas and they thrived on challenges. Their travels pushed the area of known lands further and further into the interior. If anyone would discover gold in Brazil it would be a Paulista.

Nice historians credit more likable Paulistas than Manuel de Borba Gato with the first discovery of Brazilian gold. In 1682 Gato had just successfully completed a contract for someone and as a consequence of his act was running from the law. The authorities did not take kindly to his line of work, assassination, and were in hot pursuit. Gato did what any Paulista might, he took his band into the interior, to a place on the banks of the Rio das Velhas. It was exactly the kind of place that the law could not find on their own, and if they could find it, the soldiers would elect not to go. While doing what men do when running from the law, mining, he found placer deposits of gold in the river. For a few years he and his bandeirante kept the discovery a secret. Had he found a source of copper or iron the secret would have been kept, but this was gold, and no one ever kept that kind of secret for long.

The much more historically acceptable discoverers of gold came next. When rumors of Gato's find reached Sao Paolo other Paulistas set out to this region. One of the Paulistas credited with the discovery of gold at Rio das Velhas was Antonio Rodrigues Arzao. His fame as the man who started the gold rush really rests with his brother-in-law, Bartolomeu Bueno de Siqueira. Arzao died in the wild in 1693, before he could follow up on his discovery. Siqueira, using a map supplied by Arzao, found the spot where the gold was discovered. Unfortunately another bandeirante group came by the camp while de Siqueira and his men were out hunting. This second party did some initial panning on the Siqueira/Arzao claim and was the first to make it out of jungle and into colonial civilization with tangible specimens of gold. There is every reason to believe that Siqueira did not know his claim was poached until after news of the other Paulistas' discovery reached them in the jungle. Although the second band of Paulistas did not make it into the history books their discovery did. It was their announcement that started a gold rush.

Like all gold rushes that followed, people were drawn to Brazil as if that colony were a great big sun pulling planets, asteroids, and debris in orbit about it. In 1695 men and women from all walks and positions in life gravitated to the gold strike from Rio de Janeiro and Sao Paulo. Free men left the sugar mills, priests left their flocks, merchants left their businesses without bothering to shut the doors, plantation owners left their fields and brought their slaves with them and, of course, the Paulistas came in droves and lead the way to new discoveries of gold. In 1697 the governor of Rio de Janeiro visited the mines on a royal inspection tour and stayed to work his claim with the help of retainers and slaves. He did not return to Rio until he had made himself a very rich man several years later. Right behind the colonial Brazilians came the Portuguese and behind them came the rest of Europe

The region of the first discovery became known as Minas Gerais or General Mines. In 1695 the population was little more than a few Indians and a hundred or so Paulistas. By 1709 the number of people in the area had grown to 30,000, and by 1800 the population zoomed to over 500,000. It was not easy to get to Minas Gerais. The trail started at the isolated city of Sao Paulo and stretched 20 miles across rough country to the mines. Once the gold rushers got to the diggings they found only the most rudimentary infrastructure in place. There were no laws save for local rules regulating conflicting claims to the gold diggings and river beds. Murders and thefts were adjudicated by private vengeance. In the early years food was scarce, and when it was available the price was usually an inflationary 25 to 50 times the price it commanded in the coastal towns. For the most part Brazilian agriculture was abandoned, and it took decades for the situation to normalize to anywhere near the level needed to feed all the new people flooding into Minas Gerais.

The Negro and Indian slaves were unwilling and unpaid participants in the Brazilian gold rush. They were brought to Minas Gerias when their owners, lured by the hope of easy wealth, abandoned their sugar mills. It may not have been paradise on the plantations, but be sure that the mines were sheer hell. All the heavy work was done by slaves. They carried buckets of ore by hand, sometimes they would pass the buckets down a human chain. At the head of the chain was often a rickety mine shaft which was filled with slaves working 18 hours day, seven days a week. If a slave should die he could always be replaced on the spot. No one knew how many slaves died during the early years of the gold rush and no one really cared. The gold tally was the only account that was reckoned.

Those slaves who worked the placer deposits were better off than their brothers who worked in and around the mines. It was the West African slaves who introduced the "bateia" to pan the concentrate in the rivers and streams. The first white gold rushers used their shallow tin meal plates to pan for gold; tin plates incrusted with oil and grease were only marginally useful in the recovery of placer gold. The bateia was a round wooden conical shaped bowl, anywhere from 18 inches to 2 feet in diameter. It was used in much the same manner as a gold rusher's pan, with all the concentrate sinking to a pointed bottom and adhering to a rough wooden surface.

Along with placer and underground mining, hydraulic methods were also utilized. This included the damming up of streams and rivers, digging in the dried-up water beds and using the dammed water to create jets of water that were sprayed on the river banks. The gold bearing debris was then panned or sluiced. It was not a new idea. Variations were used in the early days of mining in Spain. Another hydraulic method used during the gold rush was the cutting of a series of artificial waterfall steps into the river or stream banks. At the bottom of the terraced river bank was a trough 20 feet long and 3 feet deep. Water, upstream and at a higher elevation, was diverted from the its natural course and sent slowly cascading down the steps. This washed the soil and rock into the trough where the suspended lighter debris overflowed and ran downstream, leaving the heavier rocks and concentrate at the bottom of the trough. The slaves would regularly empty the trough and pan or sluice the gold rich material.

Gold mining was never kind to the land, but where finesse was an operational option the Brazilians and Portuguese used brute force. By using all forms of hydraulic mining they stripped the land of vegetation and valuable topsoil. A visit to the mine sites in Minas Gerais today reveals a manmade desert. Clouds of red sand and dust clog the nostrils and scratch at the throat. Where there were hills and small mountains there is now a red, flat, hard dirt crust covering dark areas of exposed bedrock. It is a hellish vision of desolation that would be more at home on the planet Mars than the dewy green hills of Earth.

Brazil was a colony and as such had to earn money for the mother country. In the case of the gold strikes Portugal wanted the same as Spain received for its mining ventures, the royal fifth or in today's terms 20 percent off the top. The Royal Crown was lucky to collect anywhere near the royal fifth during the gold rush, although it was not for lack of trying.

Before the Crown attempted to collect the royal fifth they had to enforce their right to gold. In the beginning it was the Paulistas who believed that they were the rightful owners of the mines because they made the arduous trek into unexplored lands and discovered the gold. In 1708, as more and more non-Paulista gold rushers setup claims, there was an armed struggle for possession of the mines. After losing two pitched battles the Paulistas moved on, finding gold in other areas of the interior and opening up the interior of Brazil in the process.

After putting down the Paulista rebellion the Crown put measures into effect to ensure the Crown received its full measure. First the Portuguese authorities banned any gold smelting in Brazil except at government houses. Due to poor bureaucratic planning these smelting houses were set up far away from the mines. Many a gold rusher on his way to the smelting houses had time to reflect on refining his gold smuggling methods and avoid paying the royal fifth. Another measure enacted by the Crown prohibited goldsmiths from the mining area. It was found the goldsmiths were fashioning far too many golden crucifixes. These items of worship were later melted down by the miners into ingots. In an attempt to control the flow of gold, inspection stations were set up on the important trails and roads leading out of the region. Everyone had to have a receipt showing how much gold he had, which smelting house it was bound for and where the prospector intended to pay the royal fifth. Because of this measure forgery became the leading underground industry of Minas Gerais. The battle of wits between the miners and the royal tax collectors went on until Brazil achieved full independence in 1822. So Portugal collected gold, but never the 20 percent it felt it was entitled to collect.

The boom period of the gold rush crested in 1760. Afterwards, the mines became less productive and the newly discovered gold strikes were smaller and less profitable to work. By the early 19th century the gold fever had pretty much run its course. Agriculture came back into favor with profitable crops such as cotton, chocolate and tobacco. Coffee was not an important Brazilian crop until the mid-19th century.

Gold did enter the Portuguese economy and the net result was a variation of the effect the enormous influx of gold had on the Spanish economy a century before. When gold was discovered in Brazil, Portugal was in economic decline. It seemed like a solution sent from heaven. But as the Spanish before them the Court of John V squandered the fortune. Part of the problem was the extravagant tastes of the king, who built ugly palaces and showered gifts and money on the court's favorites. The other part of the problem had wide historical consequences. In 1703 Portugal signed a treaty with England agreeing to buy English manufactured goods in exchange for the British purchasing Portuguese wine and agricultural products. With the lack of investment of Brazilian gold in Portuguese agriculture, the balance of trade quickly slipped in England's favor. To make up the growing deficit the Portuguese used Brazilian gold. It has been estimated that over one and a half billion dollars in gold was found during the Brazilian gold rush. For 50 years it looked as if there would be no end to the gold available to pay for the free spending lifestyle of royalty. Then what was thought to be impossible came to pass. The gold supply and the royal fifth declined and so did Portugal's economy, never again to rise to its former greatness.

It is said that the English industrial revolution was financed by Brazilian gold mined by Indian and African slaves. The Portuguese never held on to the Brazilian gold long enough to invest in their own industries and the industrial revolution passed them by. Many historians see the American budget deficit financed by foreign investment in much the same light. America's free spending ways based on the gold mine of easy debt may in the long run let the technological revolution pass the United States by.

NORTH CAROLINA AND GEORGIA

The founders of the North American colonies had hoped to find mineral wealth, especially gold, as Spain and Portugal did in their territories. But gold in North America remained an elusive item until two major wars were fought: one between the English and the French and one for the colonies' independence from England. There were also extraordinary political accomplishments: the establishment of the first representative democracy in North America, the first president of the United States and the adoption of an historically unparalleled constitution.

In 1799 the frontier of the newly founded country began slightly west of present day Cleveland, Ohio, bulged westward to include most of Kentucky and Tennessee, west of Nashville, then turned eastward into North Carolina, and finally south into Georgia ending at the Spanish territory of East and West Florida. In those days Cleveland was part of the Territory Northwest of the Ohio River. Ohio did not become a state until 1803. To the west of the frontier line lay The Indiana Territory, Spanish owned Territory and several smaller territories. Spain ceded the Louisiana Territory to France in 1800.

1799 was the year George Washington caught a cold and died. The same year in Cabarrus County, North Carolina, little Conrad Reed went fishing in a stream on Sunday. It was a small stream that ran through the family farm several miles from the Western Frontier Line. History records that little Conrad did not attend church that Sunday, but instead found a 17 pound yellow metallic shiny rock. His father, John, a former German mercenary, thought his son had found a very pretty rock indeed, and used it as a doorstop for a couple of years.

One day in 1802 a travelling Fayettville jeweler noticed the Reed's doorstop and purchased it for $350. The news quickly spread through the neighborhood that the 17 pound doorstop was almost pure gold. Within days excited villagers hurried to look for more doorstops in the stream. All kinds of lumps were soon found and many of the rocks turned out to be gold. The largest nugget found at this time was a 28 pound chunk. News of the find slowly traveled to more developed areas, and many of those who heard of the find rushed to the wilderness area of North Carolina to make their fortunes.

It is said that boomtowns with populations of 500 to 5,000 sprang up overnight in North Carolina. Many of these towns existed for a few years before they faded into the woodland like sandcastles on a beach. By the 1820s placer mining took second place to serious lode mining. Men from all over the world took part in the North Carolina gold rush. In addition to Americans there were workers from the British Isles, Scandinavia, Spain, Poland, Germany, Austria, Switzerland, Turkey, Italy, Portugal, Brazil and Mexico. In one mine it was reported that 13 different languages were spoken. Often this worldwide influx to North Carolina caused friction between the American and the foreign workers. Riots erupted and the state militia had to be called out to put them down. To the native North Carolinians most of the miners were considered to be a "a rough, shiftless, fighting lot." These men did not come to put down roots in the state but were always on the lookout for the next big economic opportunity.

In 1829 gold was discovered by Benjamin Parks in what was to become the town of Dahlonega, Georgia. Dahlonega is the Cherokee Indian word meaning yellow money. News first spread to the "shiftless" crowd in North Carolina and they descended on northern Georgia like a hoard of ants on a sugar bowl. Some of the lands where gold was discovered belonged to the Cherokee Indians and this was the beginning of a tragedy.

The Cherokee were a very innovative tribe of Indians and for a while benefitted from their association with white Europeans. Like all Indians of the Americas the Cherokee fell prey to European diseases brought by contact with the white man. In 1600 the Cherokee lived mostly in Tennessee and had a population of approximately 45,000; in 1650 after a smallpox epidemic the population was just over 22,000. The first official contact with the Cherokee came in 1730 when Sir Alexander Cuming, an emissary of King George II of Britain, bestowed the title of emperor on Chief Moytoy. In gratitude for receiving the title Chief Moytoy allowed his warriors to be trained by the British. Later, in 1740, under the command of Chief Kalanu, the Raven, they fought with the British against the Spanish at St. Augustine, Florida.

The Cherokee had a series of part-time chiefs who took charge only in times of war, times of peace, negotiations with other tribes and tribal ceremonies, before the white man encountered them. But the English needed one Indian to take responsibility for all negotiations between the Cherokee and British. Since this person did not exist, the British invented a sort of "treaty chief" and invested him with powers any European king would have during a normal reign. After a brief period of discord, the Cherokee accepted the notion of an overall ruler.

During the French and Indian War colonial George Washington complained to his superiors about the slowness of the Cherokee in aiding the British cause. He also noted they seemed to be an uncommonly intelligent tribe and it would be well worth it to cultivate their assistance and friendship. To this end King George III of England sent the Cherokee their first white teacher in 1765.

When the Revolutionary War broke out a majority of the Cherokee sided with the British against the colonists, even though for almost 20 years the Cherokee had fought small skirmishes with the British over territory ceded to England in 1755. The ties to Britain ran deep in the newly evolved Cherokee society. For 30 years a small number of Englishmen married into leading tribal families and some of them became chiefs. Many of the Cherokee felt the only way their children would receive a good European education was with the British. Out of over 25,000 Cherokee only 500 followed a chief named Little Carpenter and fought for the American side.

After the Revolutionary War the Cherokee still held a very favorable attitude toward the British. To counteract this potentially treasonous feeling and win over all of the Cherokee, President Washington made an offer in 1796 to all members of the tribe that they could not refuse. The United States government would use the Cherokee as an experiment in Indian education. This enterprise placed a heavy responsibility upon the Cherokee, for the success or failure of the project would determine the future of all other Indian tribes and their dealings with the federal government. To help the federal government in this endeavor, Dartmouth College, in 1799, set up loans to educate the Cherokee youth.

For nearly 30 years the Cherokee flourished. Georgia became the seat of the reorganized Cherokee nation. The Indians learned the white man's ways of building, agriculture and animal husbandry. With their new knowledge they built roads, schools, churches and log cabins. The Cherokee also adopted a republican government modeled after the United States federal government. Perhaps the most remarkable development of all was the invention of a Cherokee system of writing by a crippled warrior named Sequoya. He was uneducated and could not speak or read English, yet in 1821 he produced a workable alphabet for the Cherokee language. The entire tribe became literate in a few years and soon tribal lore and religious literature were being published. In 1828 the first Indian newspaper in the world was written in the new alphabet.

But there was gold in Georgia and much of it was on Cherokee lands. As miners rushed to the new diggings there was much friction between the gold rushers and the Indians. As early as 1802 the government of Georgia tried to push the Indians out of the state. In 1829 luck and justice had run out on the Cherokee. In return for Georgia giving up her claims to the Western lands the federal government would declare all Indian titles to the land within the state to be null and void.

Governments have bureaucracies to slow any action they don't really want to take. In 1825, 23 years after the initial declaration, the federal government finally got Chief MacIntosh of the Creeks to give up a large tract of the tribe's land in Georgia. This was all done very legal like, which means the land was surrendered by treaty. The Creek Indians rewarded MacIntosh efforts by killing him, but no matter because the next year the Creeks signed away most of their Georgia lands in yet another treaty, and this time no one was killed.

The Cherokee, sensing they would be next the Indian nation to give up their Georgia homeland, announced they would sell no more land to the whites. A war was smoldering and its flames were fanned by the gold rushers coming into Cherokee land in search of yellow money. It was unfortunate for the Cherokee that George M. Troup was governor of Georgia when the Creeks were removed from their lands. He set a precedent for the next three governors. In order to drive home the point that federal government was dragging its bureaucratic feet in the matter of Indian lands, he threatened to take Georgia out of the Union and risk war with the United States if necessary, while at the same time using his state militia to dispossess the Creeks by force. He was not reelected in 1828. The next governor, John Forsyth, a more likable fellow, followed in Troup's footsteps and persuaded the state legislature to extend the rule of Georgia over all the Cherokee lands.

At this time the chief executive of the United States was President Andrew Jackson, who was elected for his first term in 1828 on a populist platform. During the election campaign Jackson proclaimed himself to be the defender of the common man and against the special interests of the wealthy and privileged. In two terms as president his only concrete accomplishment on behalf of the common man was the seizure of Indian land, but the common man was ever so thankful.

In 1830 Jackson was able to push through Congress the Indian Removal Bill which committed the federal government to a policy of forced Indian emigration to the lands west of the Mississippi. Many of the Southern Indian nations had already accepted this principal of removal. The bill's immediate purpose was to facilitate seizing Cherokee land. Jackson's approach to the Indian question cannot be seen as typical of all frontiersmen. His most vocal opponent on the issue of Cherokee removal was the famous frontier congressman from Tennessee, Davy Crockett.

The Cherokee struggle to remain in Georgia pivoted on the 1832 U.S. Supreme Court ruling in Worcester vs. Georgia, where the court found in favor of the Cherokee against the state of Georgia and placed responsibility on the federal government to protect the Cherokee lands from encroachment by whites. The issue revolved around a Georgia law which made it illegal for whites to reside on Cherokee lands without state approval. The violators of the state law were one Samuel A. Worcester and 11 other men. These men were missionaries who not only wanted to convert Indians to Christianity, but also look out for the Cherokee interests in Georgia. Concurrent with Worcester's presence among the Cherokee, the state government was selling Cherokee land by lottery to whites without compensating the Indians living on the land. The last thing the state government wanted was outsiders interfering in this process. Jackson sidestepped the whole issue saying, "John Marshall has made his decision now let him enforce it."

What is often overlooked is what the whites were doing on the land to begin with. They had rushed into Georgia for gold. An obscure act passed in 1830 indicated the need for protection of the gold area from lawlessness and authorized the governor to "...raise and organize a guard to be employed on foot or mounted as the occasion may require to protect the mines." In defying the state law Worcester was attempting to stall the Georgia approved whites from stealing Cherokee land. Worcester failed.

In 1834 the state of Georgia allowed the whites who bought the land by lottery to come in and take over their holdings. The Cherokee were warned they had two years to relocate west of the Mississippi. In 1835, having exhausted all legal remedies, the Cherokee gave up the fight for their Georgia home. They signed a treaty ceding all their lands east of the Mississippi for $5.5. million and a larger parcel of land in the new Indian Territory west of the Mississippi. As could be imagined, there were some hold outs among the Cherokee. About 14,000 Cherokee led by Chief John Ross refused to leave until 1838. United States Army troops under General Winfield Scott were ordered to escort the last hold outs to their new home. In a five month overland journey that was called "The Trail of Tears," 4,000 Cherokee died.

Immediately after the Cherokee were driven out of Georgia, the United States government established a branch mint at Dahlonga where the Georgian gold rush began.

THE IMPORTANCE OF THE EARLY GOLD RUSHES

In a roundabout way the Brazilian gold rush contributed to the American Revolution. The Portuguese found themselves with a seemingly inexhaustible source of wealth from their Brazilian colony. Instead of investing in their own industries they bought manufactured goods from the British and eventually paid for these goods with Brazilian gold. This gold in turn further financed the English Industrial Revolution, which in turn created a need for new markets for manufactured goods. The British looked toward their colonies as an artificial market for these goods.

Trade and gold fueled the new industries and were key to the economic health of Britain, although economic subordination was not economically advantageous to the North American colonies. In order for the British to realize a profit, the colonies had to buy English manufactured goods and sell raw materials back to England or other English colonies. It would not do for the colonies to set up their own industries or to sell their raw materials to anyone that did not benefit English trade. One of the major causes of the American Revolutionary War was the establishment of economic self determination for the colonists, and the only way to accomplish that goal was by gaining political freedom from Britain at any cost, even war.

The gold rushes in North Carolina and Georgia brought foreign investments and immigrants to the United States. A good deal of the money and most of the immigrants remained in America. Some of the profits from the mining ventures were reinvested in America's own industrial revolution in the North. Some of the immigrants remained in North Carolina and Georgia, while others went West seeking more gold, more adventure or land to settle on. Of course there was always the gold, and to this end three U.S. mints were built in the South. These mints continued to operate until they were seized by the Confederacy at the outbreak of the Civil War. The need for young men in the Confederate Army put an end to gold mining in the South for the duration of the war.

Not all rewards happen at the time of maximum effort. Only a small number of gold rushers got rich from the North Carolina and Georgia gold finds. It is true that many of those who worked the fields found gold, and it is also true that the average amount of gold found for the hours put in to find and refine it amounted to little more than an excellent hourly wage, but the biggest reward came in the form of technical knowledge on how to work placer deposits. These men would have to wait a few years before they put their skills to good use, but when they did it helped to create the biggest gold rush in history.

The real losers were the Cherokee, who were displaced from their ancestral homes by waves of gold rushers. No other Southern Indian nation took to the ways of the white like the Cherokee, and no other group of Indians were so unrewarded by the whites for the effort.

CHAPTER 5

THE CALIFORNIA GOLD RUSH

California always existed, but for the longest time no one really cared. The name California comes from a 1510 Spanish story about a land of black Amazons who ruled an island "...at the right hand of the Indies, very close to that part of the Terrestrial Paradise." No one person discovered California. From 1533 to 1602 small groups of Spanish and English explorers landed their sailing ships at various points along the coast. There was not much exploration inland from the coast during this period, and California was mostly thought of as either a large island or a series of medium sized islands.

In the 16th century California was the most densely populated region north of Mexico; the inhabitants were not European but Native Americans. They were an incredibly diverse, primitive and stable people. No one knows the exact number of small Indian nations that lived in California at that time, but they had a remarkable stability. The boundaries of their tiny nations concerned land they believed was always theirs. Most tribes had creation myths which told of their people springing up from the earth where they were currently living. These convictions led the each of the Indian nations to believe their universe was unchanging; what was shall alwavs be.

One would expect to find the stability of these small nations based on richly developed agrarian communities; in fact these people were among the most primitive Native Americans ever known. With few exceptions, there was no agriculture, the only art that existed was simple basketry; where pottery was present it was the crudest in form and design, unlike most other Native Americans they had no formal governing councils. The Indians lived in paradise as simple hunters/gatherers. Food was abundant and easily obtained. The staple of the California Indians was acorn flour and seafood. When the spirit moved them, the Indians hunted for game, reptiles or birds; for variety the women would gather various kinds of nuts, fruits, grasses and herbs.

Theories on the rise of civilization usually postulate the need for leisure time within a dense population, supplying the fertile ground in which the ideas that make material progress take root and grow. It is this conceptual growth that in turn give rise to some form of civilization. In the case of the California Indians, leisure time brought about occupations centered around the pursuit of happiness. The need for material wealth in paradise seemed superfluous.

Unlike most other examples of missionary work done by the Spanish in the New World, when the missionaries came to California in the 17th century there was little disruption of the Indian way of life. Some Indians converted to Christianity, some did not. Also, unlike the confrontations in the Southwest between the

Apaches and Comanches on one side and the Spaniards who were regularly slaughtered by the Indians on the other, it was a pastoral scene of peaceful relations when the Spanish settled in California in substantial numbers. If there had been a contemporary painting showing the Spanish, the missionaries and the Indians living together its title would have been, "Having a Nice Day."

Paranoia was the reason for the Spanish to even bother settling in California at all. In the 17th century the Spanish feared the Russians or the English might grab California and threaten Mexico from the North. At the same time there was also a growing need for refitting points along the California coast for Spanish galleons sailing from Manila. So the Spanish established ports in San Diego, Monterey, San Francisco and small settlements near some of the already established missions.

The period of Spanish settlements in California was short lived. After the end of the Mexican War of Independence in 1821, California became a Mexican territory. It has been noted by some historians that had the Spanish settlers discovered California's gold, California might have become a Latin American republic or, perhaps after a short war, an English acquisition. But another nation was looking at California and many of its citizens believed they had an inalienable right to this territory.

AMERICA LOOKS WESTWARD

In the first half of 19th century America had one of the most powerful buzz words of the day was "manifest destiny." In the broadest sense manifest destiny was a concept that declared America, with its free institutions, ordained by God to create a model society in the wilderness. It was the ideological hammer used in forging a continental nation.

John L. O'Sullivan, a New York lawyer and journalist, first used the phrase manifest destiny in his own newspaper, *The United States Magazine and Democratic Review*, in July 1845. In his short article he said the future of America was "... the fulfillment of our manifest destiny to overspread the continent allotted by providence for the free development of our yearly multiplying millions."

At the time the article was written and the fateful phrase coined, the United States of America had yet to stretch from sea to shining sea. Three large parcels to the west of a north/south line running through the Rocky Mountains, the Arkansas River, the 100 degree longitude, the Red River and the Sabine River had to yet to be acquired. This boundary line was the western edge of the 1803 Louisiana Purchase, which in 1845 consisted of a huge area called the Unorganized Territory and the Iowa Territory. The three areas in question were Texas, the northern part of Mexico and the Oregon country.

James K. Polk, the 11th president of the United States, grappled with philosophical implications of manifest destiny and quickly developed a course of action consistent with the spirit of the idea. In three years from the time he took office in 1845, Polk swiftly approved the long pending annexation of Texas, endorsed the division of the Oregon country at the 49th parallel and engineered a war with Mexico in order to acquire the northern part of Mexico, which consisted of the present day states of New Mexico and California.

After the Mexican War, California was ceded to the United States in the 1848 treaty of Gualdalupe Hidalgo. News of this event took time to travel to the inland settlements. One such settlement was New Halvetia, but it would soon be known as Sutter's Fort and later the city of Sacramento.

GOLD DISCOVERED ON THE AMERICAN RIVER

Born in Germany of Swiss parents, John Augustus Sutter came to California by way of Hawaii in 1839. He had failed in every business venture he tried both in Europe and the United States. Demands for payment by his creditors kept Sutter on the move, seeking new horizons, new businesses to start and new money to borrow. He had a lively and charming personality which made it easy for him to make friends wherever he went.

Armed with letters of introduction and a self proclaimed rank of captain in the Swiss Guard of King Charles X of France, Sutter convinced the Mexican territorial governor into granting him 50,000 areas of land in the Sacramento Valley. New Halvetia was the name of Sutter's new North American empire. He built his fort at the junction of the American and Sacramento Rivers. As a stipulation of the land grant he became a Mexican citizen and in few years the Mexican government proclaimed him to be "The

▶
John Augustus Sutter. It was his mill on the American River that started the series of North American gold rushes. As many initial discoverers of gold deposits, he did not keep the lion's share of the wealth.

Hutchings' Magazine
Nov. 1857.

Marshall was a carpenter and his house was larger than most of Sutter's employees and better built.

Guardian of the Northern Frontier." What was guarding Mexico's Northern Frontier was nothing more than a large fortified homestead. "Sutter's Fort" as the locals called it did have a few towers and turrets, a palisade, a heavy gate and two brass cannons. Inside the fort there was a ragtag collection of people from all over the world: a few Belgians, Germans, an Irishman or two, and about 15 Hawaiians, including Sutter's two Hawaiian mistresses. The workers were mostly Indians who, by the standard set at the time in California, he paid generously.

Sutter maintained a large farming-ranching empire always precariously balanced between profit and debt. By now he had learned how to keep his creditors away by paying only those who threatened the loudest, but these payments were not made too often. Some creditors were never paid at all no matter how loud they yelled. The Russians only received the down payment for their fur trading post at Fort Ross. Sutter never paid a penny more for the parcel and the Russians were reluctant to sue him in a Mexican court.

His feudal domain was open to any and all travelers and immigrants who happened by, and he was generous to those who stayed for a while. One such man would carve himself a place in the annals of history; he was a 35 year old carpenter and jack of all trades, James Wilson Marshall. New Jersey born, Marshall had come to California by way of Indiana, Illinois and failed homestead in Kansas and Oregon. He was a loner who wrapped himself in a blanket of gloom. Sutter thought Marshall a bit "peculiar" because the latter believed he could occasionally see the future. Marshall's one real skill, which was far superior to anyone else's craft in New Halvetia, was carpentry.

At the high point in his career as feudal lord, Sutter needed a sawmill to provide lumber for a small city, his most grandiose dream to date. He struck a bargain with Marshall. The melancholy carpenter would go up river and select the site and supervise the building of the sawmill and

Sutter would provide the labor, tools, materials and food. Any and all profits from the mill would be split evenly between the two men. Marshall went 45 miles up the American River and found a most advantageous location, one which contained an abundance of water to power the mill and good timber within walking distance to build the proposed sawmill. There was one slight problem which went unnoticed at the time; the sawmill would be built on Coloma Indian land and not within the boundaries of Sutter's New Halvetia.

Working to build the sawmill were 10 Indians from the fort and 10 whites, six of whom were veterans of the Mormon Battalion that had gone West to fight against the Mexicans but arrived in California just as the war ended. Shortly after the Battalion was disbanded a messenger arrived from Brigham Young bearing a letter telling them to remain the winter in California, because the settlement in Salt Lake was low on rations. Most of the Battalion found temporary employment with Sutter until the following spring.

On the evening of January 23, 1848 most of the sawmill was completed. There remained a small problem; the tailrace below the wheel was found to be too shallow to turn the wheel fast enough to power the wood saw. Previously, the water was allowed to run through the tailrace to clear the loose gravel and sand. This night Marshall allowed the water to flood through the tailrace. He hoped a large rush of water would finally clear the debris.

Now, there is some confusion as to the exact sequence of events on the morning of the 24th. By Marshall's account it was a clear, cold, crisp morning, a fresh new day where the smells of ceder, pine and cypress made a fellow glad to be alive. As he walked, inspecting the tailrace, a shiny object winked at him from the bottom of a ditch filled with a foot of water. "I reached my hand down and picked it up; it made my heart thump, for I was certain it was gold. The piece was about half the size and shape of a pea. Then I saw another piece..."

A few of the workers who were there that fateful morning say that Marshall had actually seen the gold in the water the night before and waited until the next morning to retrieve the golden pea saying, "Boys, I believe I found a gold mine." To which an unknown worker replied, "No such luck." Henry Bigler, one of the former Mormon Battalion troops wrote, "This day some kind of mettle was found in the tail race that looks like gold." Bigler was the only person at Sutter's Mill who bothered to keep a diary and according to his notations Marshall found the gold in the afternoon. These minor discrepancies in the sequence of events have remained to today.

Marshall at first doubted his luck. The ugly suspicion that he had found iron pyrites began to gnaw away at this splendid moment of discovery. "...Putting one of the pieces on a hard river stone, I took another and commenced hammering it. It was soft and did not break: therefore it must be gold." By accident he had used the river's water to clear a channel and exposed the placer gold, but once he found the nugget he knew what to do to verify its authenticity. It was to be his tragedy that he never knew enough to take advantage of his find.

Three more tests were performed on the gold in the next few days. The first was done by Mrs. Wimmer, the mill crew's cook. She put a small lump of the suspect golden material in her soap kettle and boiled it with lye and baking soda. After a couple hours the nugget proved its nobility; it did not change, it was not chemically altered by the corrosive lye. The other two tests were performed by Sutter himself. Inadvertently Marshall had used one of the principles of placer mining to find his small nuggets of gold, soon those techniques would flood the region as men and women would rush to the area of the little mill on the American River to repeat Marshall's find.

Marshall finally brought his find to Sutter on the 28th of January. It was a dark and rainy day, one which could use the golden rays of great expectations. When Marshall arrived at the fort, he sought out Sutter and began speaking in a whisper, demanding they lock themselves in a room away from the servants. Sutter believed his strange partner was just being his usual peculiar self. "I only supposed, as he was queer (odd), that he took this queer way to tell me some secret." In a small dimly lit room Marshall triumphantly brought out a dirty handkerchief containing the the gold nuggets and spilled them on the table. Sutter tested the gold with nitric acid, which had no effect on the noble metal. Next he performed a test of specific density using a balancing scale and a tub of water. First Sutter put equal weights of silver and the suspect metal on two separate trays. He submerged the balanced scale in a tub of water and watched as the tray with denser gold sank. As the gold fell in the tub Sutter's dreams rose. He always believed there was gold in this bucolic land, now he was convinced. Later he would tell all who listened that he had a strange foreboding about this discovery. The next day Sutter set out for the mill to see for himself the

Monday 24th this day
some kind of mettle was
177
found in the tail race
looks like goald first discov-
ered by James Martial, the Boss of the Mill.
Sunday 30 clear & has been
all the last week our metal
has been tride and prooves to
be Goald it is thought to be
rich we have pict up more than
a hundred dollars woth last
week

February. 1848
Sun 6th the wether has been clear

Henry Bigler kept a diary in which he recorded the discovery of gold at Sutter's Mill, but he couldn't keep it a secret for long.

▲
Sutter's Mill. It was in this peaceful setting that gold was discovered and the greatest mass migration of people in North America was launched.

extent of the find on the American River.

Not content to let events run their own course, Marshall had some of the mill workers seed what gold was already found in the tailrace. Fate can seem whimsical in its ways. The idea was for Sutter to naturally find the gold just laying on the bedrock. Salting a mine, as the practice was called, would soon become a major Californian industry among con men seeking investors in worthless land. The state's first recorded case of this type of fraud would backfire and yet succeed. Before Sutter arrived at the sawmill one of Mrs. Wimmer's young sons found the salted gold in the tailrace and ran excitedly toward the workers camp with his pocket full of the small shiny pebbles. On the way he met Marshall and Sutter and showed them his find. The boy was so earnest in his belief that he had found the gold just lying in the tailrace, he immediately convinced Sutter of its authenticity. Sutter examined the small gold nuggets and exclaimed, "By Jo, it is rich!"

Now Sutter had two tasks to accomplish: the first was to obtain title to this land and the second was to keep the find a secret as long as possible. To implement the former Sutter signed a three year lease with local Indians for all the land within 12 square miles of the mill; in return he would pay the Indians an annual fee in clothing, tools and food. To execute the latter Sutter asked his workers at the mill to keep the find quiet for six weeks until the work on the mill could be completed. In truth he wanted to buy time until he could gain clear title to the land.

IT BEGAN AS A TRICKLE

Marshall's find at Sutter's mill was not the first gold strike in California. In 1842 a small group of Mexicans discovered gold around the Los Angeles area. For whatever the reason, perhaps the United States' war with Mexico, the find became a tiny footnote in history. Some historians believe gold was also found at the same time in the Sacramento Valley, but again it did not create a wave of gold seekers rushing to California. There appeared to be little or no interest at that time in a few random gold strikes in northern

Mexico. It would take a far more believable person than a few unknown Mexican miners to spark a gold rush.

The first gold rusher was Henry Bigler, the diary keeper. As those who were in the know were busy telling everyone else at the Sutter's Mill camp about the secret gold find, Henry was busy prospecting on his days off. When asked where he was going, Bigler told his fellow workers hunting. Indeed he was hunting, not game or pheasant but gold, and his hunting ground was not near the sawmill but in the surrounding streams and gullies. On February 22, Washington's Birthday, Bigler counted out his pickings and estimated it amounted to almost $40.

Judging from all his clandestine prospecting one might assume that Henry Bigler could keep a secret, but he could not. Shortly after counting his gold Bigler went to visit former army mates from the Mormon Battalion. They were building a gristmill for Sutter a few miles from the fort. They heard about his frequent hunting expeditions and grew suspicious. Henry was not one who went hunting often, least not the Henry Bigler they knew. Upon being confronted by his friends for a more believable explanation of his long absences, Henry unrolled his knotted shirt tail and took out an ounce of gold dust; the amount represented a week's diggings. Now the Mormon carpenters at the gristmill were finding it very hard to concentrate on their work. The secret was certainly making the rounds of New Halvetia.

Sutter believed Mexico would cede California to the United States in the very near future. To insure his title to land he sent one of his workers, Charles Bennett, as his emissary to the United States military governor in the provincial capital at Monterey. Bennett would not do any negotiations on Sutter's behalf, instead he carried a very carefully worded letter to Col. Richard B. Mason. Sutter omitted any mention of the gold find from the letter, rather, he concocted a story about the need to protect the foothill Indians from the more aggressive mountain tribes. If he had title to the land surrounding the sawmill he could better accomplish this peacekeeping duty.

About the time that Bennett left for Monterey Mrs. Wimmer, told a teamster, Jacob Wittmer, about the gold find. At first Wittmer thought this to be idle gossip cooked up for the enjoyment of bored mill workers. But her young son, the curious boy who picked the salted gold from the tailrace, proudly displayed his find from the mill. In fact, the family had done some more prospecting and had quite a small gold stash of their own. For whatever the reason, Mrs. Wimmer gave Wittmer some of the gold nuggets. Perhaps she was very generous woman or perhaps there was no Mr. Wimmer at this time, nevertheless another person had a story to tell and some gold to show as proof.

As Mrs. Wimmer was giving away some of her gold, Bennett was showing his. He made an overnight stop in a little village of Benicia. There, in a small country store, Bennett heard several men talking about a recent coal find. Unable to contain himself, he made some disparaging remarks about black rock mining. Having gained their attention, he told his listeners there was a easier and more important mineral to mine, gold! To punctuate his statement he showed them his own bag filled with gold dust and pebbles. Bennett also told another group of men at another overnight stop in a San Francisco hotel. This time one of the men was a former Georgia gold rusher, Isaac Humphries. He would soon go to the diggings and stake a claim; he would also introduce the rocker to California and to all subsequent Western gold rushes to follow. Such was the influence of the all but forgotten Georgia rush. Bennett was enticed by Humphries to show him some proof of this gold strike. Again the bag of gold came out of its hiding place. But this time the man examining the gold knew more about the subject than anyone else for hundreds of miles around. One look was all it took to convince Humphries that this was going to be a richer gold strike than all of Georgia.

The conclusion of Bennett's trip to Monterey was anticlimactic. Upon arriving he promptly called upon Col. Mason and presented Sutter's letter. Mason read the letter and the planned lease of the land and dismissed the whole proposal, declaring the United States government does not recognize the right of Indians to sell or lease land. Then, as before, the talk turned to the subject of gold and Bennett once more produced his little bag of gold for Mason to see. Mason called in his adjutant, a 28 year lieutenant, William Tecumseh Sherman. After testing the largest nugget by flattening it with the back of a hatchet, Sherman told his superior that the sample was definitely gold. Bennett returned home and left the two army officers to ponder how big the gold strike might be.

The golden cat was creeping out of the bag of secrecy and beginning to climb the wall. For a short while Sutter's workers performed their assigned tasks and went gold prospecting on Sundays. But as projects were completed more and more workers went out and stayed in the streams and rivers to pan for gold. As soon as Henry

Bigler's fellow Mormons completed building the gristmill they left the fort and went up the American River to the sawmill to have a look for themselves. They dug around for a few days and discovered enough gold to convince them to become prospectors. Halfway back to the fort the Mormons noticed certain similarities between a previously unnoticed sandbar and the riverbed near the sawmill. They began to dig and pan on the sandbar and were rewarded with a sizable gold find. When Bigler visited his friends in March, they were taking $250 a day from the bar. By then everyone called the sandbar Mormon Island.

Sutter's dream of New Halvetia was becoming a nightmare. The empire he envisioned was crumbling. Throughout the fort workers were deserting at the rate of two and three a day. News of the gold strike was spreading by word of mouth all over the Sacramento valley, not an efficient method of communications. On March 15 the San Francisco *Californian* ran a small article titled: GOLD MINE FOUND. In a low-key manner it described Sutter's Mill and included a few details about how much gold was found. According to the article, "One man brought thirty dollars worth to New Halvetia, gathered there in a short time." A rival newspaper, the *California Star*, published a similar story the next day. Instead jumping at the chance to make a fortune, the 800 or so residents of sleepy San Francisco didn't even bat an eyelash at the news. Many believed it was old Capt. Sutter trying to pull off another credit scam.

What began as a presidio, or Spanish fortress, later became a mission and finally a small pueblo. San Francisco, which takes its name from the mission which was dedicated to St. Francis of Assisi, was a tiny Mexican trading port called Yerba Buena when the U.S.S. *Portsmouth* dropped anchor at Yerba Buena Cove in 1846. After a few days of waiting with its guns trained on the small port, Capt. John B. Montgomery received orders to take possession of the town and announce to its citizens that the United States had annexed all of California.

Within two years of American occupation several large scale changes were made: streets finally had names, the population grew from 300 to 800 people, buildings increased from 147 to almost 500 and shipping became a major industry with 86 ships using the port during the year ending March 30, 1848, a few weeks before gold fever seized the townsfolk. Rumors of gold flooded San Francisco and if most of the population ignored the stories of gold strikes near Sutter's Mill, they created a climate for the right incident to exploit.

A SMALL FLOOD

Sam Brannan was the right man at the right place with the proper amount of credibility to send a flood of San Franciscans rushing to the gold mines. His birth place was in Maine. He became a Mormon while running his own printing shop in New York City. To escape persecution in 1846 a large number of Mormons left New York state for a home beyond the frontier. A year later Brannan took a small contingent of Mormons aboard the sailing ship *Brooklyn* and headed West. Their trip took them around Cape Horn into the Pacific Ocean, with the last stop being San Francisco. Brannan hoped this sleepy little town would become a Mormon paradise; he was very disappointed to hear the unwelcome news that California was under United States rule. However, the flamboyant, energetic and often unpredictable Brannan never lost his stride. No sooner did he disembark from the *Brooklyn* than he went about setting himself up in several business.

He was considered to be the leader of the California Mormons, although most people in San Francisco did not know he had been excommunicated by Brigham Young in the spring of 1847. They had a fierce clash over the issue of where the Mormons should make their future home. In the ensuing debate over the site of the Promised Land, Brannan campaigned for California and especially San Francisco. Young wanted to settle in the Salt Lake Valley, Utah. Young believed that San Francisco was a city of sin and his followers would easily go astray in the face of such overwhelming temptation. San Francisco suited Brannan just fine and after losing the argument and being drummed out of the Church of the Latter Day Saints, he left Young near Fort Bridger, Wyoming and returned to California.

Brannan founded and owned California's first newspaper, the San Francisco *Star*, the very same paper that ran the second gold discovery article. His fingers touched every economic and political pie in San Francisco. One of the more profitable ventures was a half interest in a trading post at Sutter's Fort. It was at this trading post that the teamster, Jacob Wittmer, appeared with some gold nuggets in his possession. One of the shiny pebbles came from Mrs. Wimmer, but the others he found in the river. Wittmer made a number of purchases, however, the largest item on the bill was for whiskey. Brannan's partner was the proprietor, George Smith. He haggled with Wittmer over the value of the small gold nuggets and settled on a price of $8, just enough to pay for the goods with a little left over for change. Sensing this was an unusual transaction, Smith sent the gold to Brannan in San Francisco.

This was too much for Brannan. He had sent the editor and a reporter up to the mill and they found nothing out of the ordinary there. With 20/20 hindsight this was natural. The workers at the mill were following Sutter's request and not telling outsiders about the gold. Also, they also did not want the already crowded area becoming overrun with new prospectors. Brannan, with gold nuggets in his hand and more than a suspicion on his mind, went to Sutter's Mill.

During the first two weeks in May Brannan poked around the Fort, the Mill and all the rivers and steams nearby. On his return in mid-May he strode down Montgomery Street holding a small quinine bottle filled with gold nuggets high in the air for all to see. He looked a little mad with his black hair uncombed and his clothes soiled with trail dust as he shouted at the top of his lungs, "Gold! Gold! Gold! on the American River!" That was all it took to virtually sweep San Francisco of every able-bodied man, woman and child and send them rushing up to the mines on the American River.

The first gold rush summer and autumn in California were like a dream. If shown in a movie, the idyllic time would be shot in gauzy soft focus. It may have been the only find where the gold was so accessible. In his report to President Polk, Col. Mason said, "No capital is required to obtain this gold, as the laboring man wants nothing but his pick and shovel and tin pan with which to dig and wash the gravel."

At the start of the 1848 rush many of the would be miners from San Francisco were dismayed to learn they actually had to work to get at the gold, even though at many of the new sites near the American River some of it could be picked off the top of stream and creek beds. If there was no gold to be found in top sand or gravel the next place to look was in the stream's crevices. Some of the early miners would strike it rich just using an ordinary pen knife to pry the gold nuggets and flakes out of the rocky fissures. If all the easy methods of gold gathering failed, the gold rushers had to resort to hard work using the pan and rocker to uncover the golden fortune they were so desperately pursuing.

The availability of the gold was due to the erosion of quartz locked gold veins by glacial and climatic forces. The larger pieces of abraded gold lodged behind boulders and and in cracks in the bedrock of rivers. The smaller flakes were carried downstream until they mixed with the gravel or fell out of the current in quiet eddies. Unlike any other gold rush past and future the gold found at Coloma was unequaled in the ease of discovery and the large quantities easily mined.

Everyone wanted to find gold, so almost everyone in California went into the hills to try their luck. In the first few months of the rush blacksmiths, teachers, herdsmen, sailors, and farmers quite literally dropped what they were doing and went searching for gold. A captain of a sailing ship, flying the Peruvian flag, anchored off of San Francisco cove and waited for the usual small boats to come and greet him. He grew worried when none showed up and found it even stranger when he could not see anyone on the beach. The captain went ashore and walked for about a half an hour along the deserted San Francisco streets before he met a man who enlightened him as to the cause of this curious state of affairs. Needless to say the captain had a difficult time keeping his own sailors from jumping ship and joining the gold rushers in the hills.

Soldiers were not immune to the gold fever. The pay for an army private was $7 a month. Inflation, due to the shortage of goods and labor in the gold country, meant a soldier's meager wages could just about buy three pounds of flour. In a letter to the War Department he noted, "The struggle between the right and $7 a month and wrong and $75 a month is severe one." Col. Mason's troops began to desert. The military governor tried sending soldiers to pursue the deserters only to find that the pursuers became the pursued as they too went off to the mines. Things were so bad in August that when Commodore Jones, of the U.S. Pacific Squadron, Col. Mason and lieutenant Sherman had dinner in Monterey, they had to prepare the evening meal themselves since all the cooks and servants had run off to the gold mines. Often Col. Mason gave the missing soldiers leave after the fact, so they would not be officially classified as deserters.

The gold rush spread like a red wine stain on a white linen tablecloth. The miners worked their way north and south to the other streams flowing from the Sierra Nevada. Some gold was found in all of them. The 10 man party of Gould Buffum was typical of a successful gold mining expedition in the idyllic season of 1848. Buffum and his men started out working a claim at Weaver's Creek, which yielded $100 per man per week. Yet Buffum thought this was a mediocre find and the party moved to the Middle Fork of the American River, where the take increased to $40 per day per man. The claim stayed at this level productivity for a month.

Around August miners north of San Francisco suspected that somebody somewhere was finding even more gold than they were. This was a case of greed coupled with the ease in which the gold was found. At a camp named Dry Diggings, the future site of Placerville, the miners who were dry washing $50 a day in June and July saw their daily take dwindle to under $35 a day. When news of a new gold strike to the south near the Stanislaus River arrived in the camp over 200 men hastened over grass covered foothills to the richer find.

As gold rushers were scampering to the Stanislaus River, a party of 30 Mexicans led by Antonio Francisco Coronel, a school teacher and future mayor of Los Angeles, were on their way to the newly deserted Dry Diggings. With a few days journey to his goal, Coronel met a priest carrying a bag of gold. The priest told the company of the richer find on the Stanislaus River and Coronel changed direction toward the river. That night while camped in a deep chasm he was confronted by a small band of Indians. It was obvious to the Mexicans that the Indians wanted their saddle blankets and serapes, they wanted them real bad. Through skillful negotiations Coronel learned they had gold. Eventually the Indians paid about a pound of gold for several blankets.

Coronel knew an opportunity when he saw it and he knew how to take advantage of it. He had his best scout follow the Indians as they left his camp that evening. The scout tracked them for a day until the Indians made camp in another large ravine. He watched the Indians digging in the gully and after they rode away the next day he tried the ravine with a pen knife. In less than an hour he dug up 3 ounces. He dashed back to Coronel and told him of the find. Coronel went to the sight with a few of his men and in no time was averaging $500 a day. Such were the seeming limitless opportunities in that first gold rush year.

As men went to the mines a severe labor shortage developed. Towns such as San Francisco and Monterey were all but abandoned. There was no one left to do the ordinary tasks that keep a robust economy running. It takes people to grow food, unload ships, keep the shops and repair all things great and small. An early history of San Francisco quotes eyewitness accounts as to the state of the city in June of 1848. "...The prices of labor and all necessities rose exceedingly. The common laborer, who had formerly been content with his dollar a day, now proudly refused ten; the mechanic, who had recently been glad to receive two dollars now rejected twenty for his days services." Merchants were in the enviable position of being able to charge almost any price for needed items at the gold sites and at the same time being in the desperate position of having to deliver the goods themselves.

Inflation rose so fast that summer it would make present day third world countries wince. Gould Buffum's 10 man party paid $300 for their supplies in late July. By early October the same supplies cost over $1,000. Coffee, for which he paid less than $1 a pound, rose to over $4 when he restocked his supplies. In December a breakfast at Sutter's Mill of sardines, bread, cheese, and beer cost $43. Those not busy finding gold were busy making money.

When the rush started Sam Brannan opened up a general store near the mill. It was stocked with provisions such as food, picks, shovels and iron pans. The last item was used, of course, to pan gold, and Brannan had bought all the available pans at the beginning of the San Francisco rush for 20¢ each and sold them from anywhere between $8 and $16 apiece, payable in gold. Brannan's other venture, the store at Sutter's Fort, did very well too; it grossed $36,000 in the first four months of the gold rush.

The first year paradise of the Californian gold rush was not without its rotten apples. All sorts of diseases flourished in some of the mining camps. Because the camps appeared in very much of an ad hoc fashion there were no sanitary facilities. Miners not wanting to lose a minute of daylight sometimes neglected to build even slit trenches for their waste matter. Illnesses arising from this situation were common. Outbreaks of dysentery and cholera were reported in some settlements. Due to the lack of fresh fruit and vegetables scurvy, a vitamin C deficiency disease, became a commonplace ailment, even though it was known as a sailor's disease. Although it was known that eating lemons and limes could cure the illness, there was no one to deliver the produce to the gold fields, and many miners suffered from bleeding gums, lost teeth, hemorrhages, exhaustion, diarrhea, profound weakness and sometimes death. The combination of these illnesses in any given settlement gave rise to a fetid odor which could be detected over a mile away.

Even when there were no major outbreaks of disease, gold rushers often fell ill from neglecting minor colds which grew worse and became full blown cases of pneumonia. When physicians went to treat patients at the gold finds they themselves caught the gold fever and ended up digging for gold. Ignoring their own heath, the doctors would fall ill to the same ailments they came to treat.

If idle hands make work for the devil, then the devil was vastly disappointed in the gold rush of 1848. With the exception of the rustling of Sutter's cattle and one bad murder involving several sailors, one of the great wonders in the first year was the absence of crime at the various digging sites. All the miners had the desire to do was to work their claims and find gold. At the end of a work weary day they did not have enough energy left in their stream chilled bones to get into trouble carousing at night.

Col. Mason, in his role as military governor, allowed all who wished, to work their informal claims freely and without governmental interference. Two factors went into his decision: American land policy at this time was favorable to individual initiative and Mason had no choice since he lacked the manpower to enforce the federal government's rights to the land. Sending any of his troops into the mining sites would be seen as an open invitation to desert. A few men made a large fortune that year. For the average gold rusher the take was about $20 a day. Not a bad wage when you consider that laborers in the East were earning $1 a day, when work was available.

In 1848 the system of claims was as follows: anyone who cleared the topsoil from any portion of the bar of a river was recognized as the claimant of said parcel of land. When two men differed as to the ownership of claim they would ask a third to act as judge. In some mining communities the disputed claim was decided by an impromptu meeting of the miners. The loser accepted the verdict and went on his way; there was always gold to be found in the next valley. Everyone believed that gold was plentiful, all one had to do is find the right claim. Theft was made obsolete in the face of so much wealth lying just below the surface of the Californian rivers and streams. Because of this prevailing belief bags of gold were often left unguarded near the miners' claims. Years later some of the 1848 miners would bitterly say, "We needed no law until the lawyers came." And they would come with the hoards the next year.

Gould Buffum told of an incident that occurred in January 1849 which foreshadowed the coming of the forty-niners and the rise of crime at the mine sites. It seems that three men, one Chilean and two Frenchmen, were caught red handed attempting to rob and murder a gold rusher near the mining community of Dry Diggings. The three men were beaten senseless at the scene of the crime and brought to a nearby house while the miners deliberated on the fate of the felons. One of the miners became the judge and about 200 miners became the jury. It was agreed that the alleged perpetrators were guilty of attempting to commit a crime and not the crime itself. But these three men were anything but popular with the gold rushers at Dry Diggings. The crowd felt the would be thieves were guilty of some unknown crimes and it would be best for all concerned if they were disposed of. After a 30 minute trial the trio was found guilty of attempted murder and robbery.

A certain excitement ran through the crowd; it was mood fed by the process of the trial and the imbibing by the miners of a substantial amount of liquor. In this atmosphere of mob comradery anything seems possible. When someone asked, "What punishment shall be inflicted?" The answer came from a drunken brute of a miner who said, "Hang them!" And the intoxicated miners began chanting, "Hang them." And the chant became a chorus united in a single purpose; they became a creature known as the Juiced Justice of the Gold Rush.

According to Buffum he was practically the only sober miner at the trial. He mounted a tree stump and spoke, in the name of God, humanity and law, against the proposed drastic action of vigilante execution. But the mob was not interested. Buffum said "They would listen to nothing contrary to their brutal desires, and even threatened to hang me if I did not immediately desist from any further remarks." And so the the condemned men were led out of the shack they were held in and marched to a spot where three ropes were attached to the limb of a tree. They cried for mercy when they were placed on a wagon, they wept, they made short speeches in their native tongues that none of the miners understood–and they were hanged. It was this incident that gave the mining community of Dry Diggings the new and notorious name of Hangtown. The name Hangtown persisted for a couple of years until it was changed in 1851 to the more sedate Placerville. The town also became known for its "Hangtown Fry" which was accidently invented in 1849 and is still served today.

As the story goes a hungry and gold laden miner, fresh from his claim, walked into a wood framed hotel called the Cary House. Feeling momentarily prosperous he asked the waiter for the most expensive meal the cook could make. The waiter replied that would be oysters, eggs was the next dearest item on the menu. Thinking for a moment the miner said, "Fry a mess of both and throw in some bacon." A rough and ready recipe

▲
Throughout the 1848 phase of the Californian gold rush conditions were idyllic. For those who panned the streams the first year the weather was right, crime was none existent and if one area had too little gold the next valley would definitely be better. With the coming of the Argonauts all this would change for the worse; it would become another example of "Paradise Lost."

was born from the hurly-burly days of the gold rush.

A TIDAL WAVE

In 1848 word of a breaking news story moved much quicker than in the time of the Brazilian gold rush, but much slower than we are accustomed to 140 years later. Col. Mason sent a report about the gold rush at Sutter's Mill on August 17. It did not arrive in Washington, D.C. until December 9, although word of the discovery preceded it by telegraph. In the 1840s everything moved at a slow pace. Even a gold rush had to wait its turn.

It took 25 uncomfortable days to travel from the Mississippi River to the Pacific by stagecoach, and this was pushing it. Outside the stagecoach was brightly colored and decorated, but inside it was crowded and dark. Some stagecoaches had three parallel rows of planks for benches, each seating three passengers. Only the last plank, reserved for women, had a back rest. In the smaller two plank model the passengers sat facing each other, and the rear plank or bench had a backrest. Four horses pulled the stagecoach over bumpy roads and inside the passengers were often tossed about in the small compartment; springs were unknown at this time and therefore there was no suspension to absorb the shock of iron clad wheels hitting ruts and rocks in the road. The only protection from bad weather was leather curtains which could be laced closed when it rained. Travelers often drank themselves into a stupor for the entire trip. Occasionally a passenger suffered from sleep deprivation, went mad and had to be tied to his seat for the remainder of the trip. All this for the most expensive form of travel.

The telegraph was in operation at this time only from the East Coast to New Orleans on the Gulf Coast. Railroads frequently traveled between the East Coast and the Mississippi River in 1848, but the Atlantic and Pacific Oceans would not be linked until 1869 with the completion of the Union Pacific. If a message had to reach across the continent it was hand carried across the continent, or by ship around Cape Horn, or it went by mail.

Lt. Lucian Loeser was intrusted by Col. Mason to bring his report and tea caddy filled with a little over 230 ounces of gold to Washington. To accomplish this task Loeser had to take a sailing ship to Peru where he boarded a British steamer bound for Panama, whereupon he traveled by land across the Isthmus and took another ship for New Orleans, where he booked a seat on an express stagecoach to the capitol. It took just about five months to make this momentous trip and this was really moving by the standards of the day.

Polk was a lame duck president who accomplished a great deal in his one term. Now he was to change history again. He presided over the annexation of Texas, the acquisition of the Oregon Country, the successful completion of a war with Mexico and the annexation of the northern part of Mexico. Now Polk was looking for one last achievement to cap off the last days of his term. The opportunity came with Mason's report; it was news rich with promise, it was the discovery of gold in California.

In his December 5 opening message of the second session to the 30th Congress Polk said, "It was known that mines of precious metals existed to a considerable extent in California at the time of its acquisition. Recent discoveries render it probable that these mines are more extensive and valuable than was anticipated. The accounts of the abundance of gold in that territory are of such an extraordinary character as would scarcely command belief were they not corroborated by the authentic reports of officers in the public service." Two days later Lt. Lucian Loeser arrived in Washington with Col. Mason's report and a tea caddy filled with a little over 230 ounces, or $3,600 worth, of gold. It was immediately put on display in the War Department. News of the Sutter's Mill gold strike had reached the East before December 1848 but no one would take the stories seriously. It took the trusted word of the President of the United States to turn a California rumor into a nationwide tidal wave of gold seekers rushing to gold laden rivers of the West.

Col. Mason's report also included an estimate of 4,000 men working the gold finds. By the end of the year men from Oregon, Mexico, Peru and Hawaii migrated to the gold fields and raised the number to 10,000. As it turned out, this was only the beginning. This was a young country, a nation which had recently expanded its boundaries by adding large parcels of land in the West. A nation which was undergoing revolutions in industry and technology, a nation where new inventions like the telegraph, the railroad and steam engine were changing everyday life.

One fact often overlooked in the histories of the Californian gold rush is the age of the participants. They were all young and almost all male. Pictures of them taken at the time show bearded men with long, sometimes matted hair cascading out of weathered felt hats. What is hidden from view is the youth of the miners.

Horace Greeley, American newspaper editor and political leader, passionate advocate of the westward movement and author of the famous exhortation, "Go West, young man!" owned and edited the *New York Tribune*, a newspaper dedicated to reform, economic progress and the elevation of its readers ...the masses. On December 9 he caught a bad case of gold fever and wrote an ed-

Images of the Gold Rush

I

Once Horace Greeley endorsed the gold strike at Gregory Gulch new Argonauts were only too willing to follow his advice and go West to the new diggings.

Gold and silver in bar form. These were the basic forms used, after refining, for the convenient and standardized handling of precious metals. ▲ ▶

2937
2938
1026 80

▲
First gold coin minted at the United States Assay Office.

▲
Before there was an official government minting office in San Francisco, miners would have private assay companies stamp their gold into coins for convenience when travelling and spending. This is one of the first coins to be stamped.

▶
In San Francisco such a surplus of ships existed that many were used as stores and hotels. In this mid-century lithograph one is anchored bow-on between a saloon and general store.

NIANTIC HOTEL.
LITHOGRAPHY AND PRINTING
BUBB GRUB &CO
NOTARY PUBLIC
COL TIBBS DENTIST
GENERAL STORE
BOGGS
LIQUOR STORE
TRACT SOCIETY.

▶
Mud Lake Basin. This steep trail was the end of the line for many wagons. They couldn't make it down unless slowed by double-locked wheels.

▶
Charles Nahl painted Sunday at the Mines from memory, years after his 1850 visit. The left half of the painting depicted activities which are best described as a continuation of the Saturday night follies. The right half of the painting shows rather quiet endeavors, which were more in keeping with the heavily religious feelings at the time.

Miners are at rest. In a primitive cabin the miners cook, drink, sleep and, most importantly, weigh the day's take of gold.

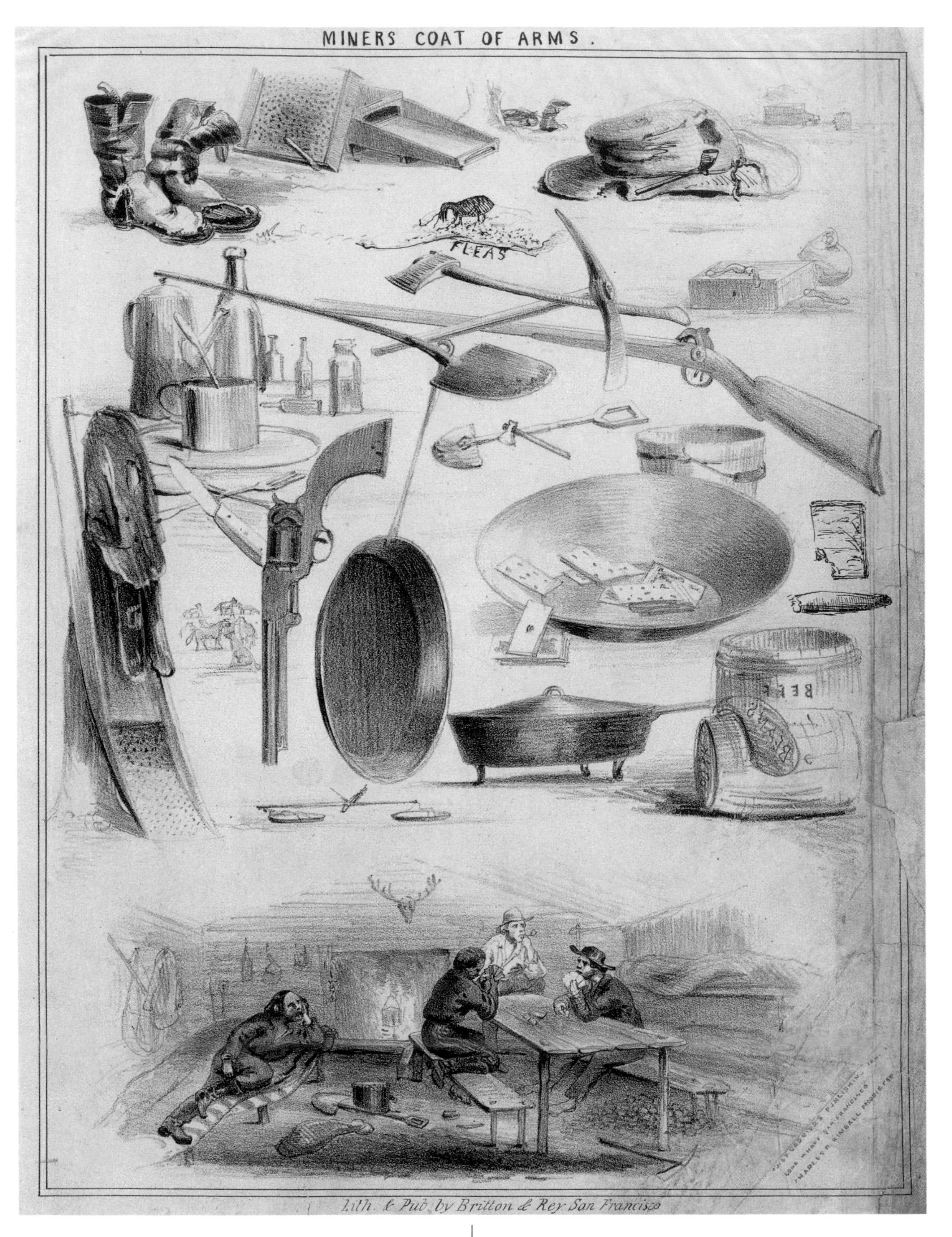

Pictured are everyday items in the life of a placer miner. The pan and rocker were universal tools of the trade. The beans, pork and bully beef were the mundane sustenance that kept the work going, while whiskey, coffee, chewing tobacco and playing cards gave the miner a chance to forget his aching muscles and constant flea bites. Firearms place this picture in the 1850s time period because they were not a miner's necessary companion until claims became scarce.

itorial which stated, "We are on the brink of the golden age. We look for an addition within the next few years equal to One Thousand Million Dollars to the general aggregate of gold in circulation." On December 11 Greeley's gold fever worsened and he wrote, "We don't see any links of probability missing in the golden chain by which Hope is drawing her thousands of disciples to the new El Dorado, where fortune is as plentiful as mud in our streets."

In the nation's capital a reporter from the *Washington Daily Union* was so overwhelmed by the amount of gold in Col. Mason's tea caddy that he wrote, "...The gold in California must be greater than has hitherto been discovered in the old and new continents." Other newspapers spread the golden word throughout the Union. The skeptics of a few months earlier caught the fever and began looking for gold seeking equipment and prospecting advice.

It has been noted that our contemporary society seems to be run on the principles of "hype" and "scam." Everything new is promoted to unbelievable heights through the mass media of television. Salespersons invade our homes using the latest in telemarketing techniques. In both cases the product may turn out to be much less useful or fulfilling than promised. The product or service bought is just an imaginative variation on the theme of greed. This complaint is not unique to our age. It is a timeless human foible, the willingness to believe in the unbelievable even though common sense would advise otherwise. So it was during the early stages of the California gold rush. Would be miners began purchasing items advertised as essential for success in the quest for gold. Some of the products such as tents, India-rubber wading boots, revolvers, money belts, shovels, medicines, cannons and rough mining clothing seemed reasonable to the Eastern man ready to risk everything on the possibility of becoming wealthier than his grandest dreams. Pawnbrokers were only too happy to accept watches, rings, musical instruments and anything else imaginable that could be hocked to pay for the passage to the West. Other entrepreneurs fed these dreams with promises of instant success for miners if they would only buy their: "Chrysolyte California Gold Finder," "Archimedes Gold Washing Machine" and any number of other useless contraptions.

A new industry sprang up overnight. Its sole concern was the profitable feeding of the national gold fever. Guide books, often inaccurate, were churned out by hack writers. A good number of these 19th century instant books were rehashes of John C. Frémont's famous *Report of the Exploring Expedition to the Rocky Mountains in the Year 1842 and to Oregon and North California in the Years 1843-44* and Landford Hastings' *Emigrants' Guide*. Others were fanciful elaborations of Mason's report, and still others were mostly fiction masquerading as fact.

If incipient gold rushers were not content to read all they needed to know they could also attend lectures such as "On the Structure and Formation of Rocks in Which Precious Metals Could Be Found," given by enterprising geologists. These talks were, more often than not, popularized college lectures that the instructors had given before less enthusiastic students of the stony science. But with each passing week the lecturers' claims of the amount of gold to be found in California grew by leaps and bounds. Audiences heard of miners who were making $1,000 a day and Indians making over $3,000 a day. When a humorist told of a miner who sat on an 839 pound gold nugget for two months in order to prevent anyone from stealing his find, the audience failed to see the joke and took the whole story seriously. It was the willingness to believe in the California El Dorado that put the nation's common sense to sleep.

Perhaps it was the tone of the times. Interest in spiritualism had been growing in America almost as fast as the interest in gold. Typical of this phenomena were the experiences of two sisters, Margaret and Katherine Fox of Hydesville, New York who, in 1848, heard mysterious rappings at night, "...as though someone was knocking on the floor and moving chairs." Several doctors who witnessed the event said the rappings came from the sisters cracking their knees. The sisters claimed the rappings were communications from the spirit world. Despite the dissenting medical opinion, the sisters were in demand by everyone, including such notables as James Fenimore Cooper and Horace Greeley who avidly attended their séances. In this atmosphere of gullibility all sorts of occult practitioners came out of the woodwork to lend a hand to those in need.

Astrologers were in demand. Many in the 19th century, as today, believed their eventual success could be foretold in the stars; if they did such and such on a certain day then such and such would surely follow as a consequence. If leaving New York the first of February meant a gold rusher would become a millionaire and leaving there on the 15th meant he would only find $100,000, then by all means he should leave on the first; for some reason astrologers felt the sooner their client left

the better. History does not record which occult studies certain clairvoyants used to convince their patrons to take them along to California. Nor is it recorded whether the clairvoyants stayed with their clients at the gold fields or went searching on their own. All that is recorded is they too found a way to seek their fortunes. History would call them the "forty niners," but in their day they were known as the Argonauts, after the party of adventurers in the Greek myth who sailed with Jason aboard the *Argo* searching for the Golden Fleece. Much like the Greek voyagers of yore, many of the latter-day Argonauts banded together by forming companies and associations.

The companies sometimes were as small as 15 men or as large as 250. They came in all shapes and sizes. There were all German companies as well as those composed of all Harvard graduates. One was made up of all Cherokee Indians and one was founded by the retired matron of Sing Sing prison. It was named the California Association of Women and included one maiden, two widows and 15 men. The companies' officers were elected, but in every other respect they were very businesslike. Some associations took on a military look, complete with the wearing of uniforms, the blowing of bugles and occasional martial drills complete with bands. Whatever the outward affectation, shares were equally allotted between the officers and the privates. Since one in six Argonauts came from New England, it is not surprising that the largest contingent of associations would also hail from New England. Edward Everett, president of Harvard, was quoted in the Boston newspapers as urging his neighbors to, "Take your Bible in one hand and your New England civilization in the other and make your mark on that (California) country."

There were three ways to get to California from the East Coast: one route was by sea to South America, around Cape Horn to the Pacific and up to California; another route was a combination of sea and land travel, which meant sailing to either Panama or Nicaragua and trekking cross country to the Pacific, where the gold rushers boarded a ship bound for California; the last route was overland across the Western wilderness by means of several trails. In the winter of the 1849 the land passage to California was snow covered and frozen. The vast majority of poor but

◀
Established early during the California gold rush, hanging was the penalty for dastardly crimes. As many who came before him, McCall was hanged for murdering Hickock.

adventurous gold seekers had to wait until after the spring thaw in the West. But for those who had the money the sea routes beckoned and could be traveled immediately.

The advertised fare around the Horn was $500, but frequently the actually amount paid was over $1,000. A passenger could expect the average trip to take six months although, depending on the winds and the weather, the journey could be as short as four months or as long as eight. The fastest ships were the freight only clipper ships which regularly made the Boston to San Francisco run in 100 days. The record of 89 days for the trip around Cape Horn was held for 135 years by the clipper ship *Flying Cloud*. On February 12, 1989 *Thursday's Child* broke the gold rush record making the journey in just under 81 days.

Taking large amounts of baggage was a big advantage of the fast freight clipper route to San Francisco. This method of travel was preferable for those gold seekers who felt that they could not survive without all the comforts they were accustomed to in the East. Enos Christman, a young printer's apprentice from West Chester, Pennsylvania, was such a traveler. According to Christman the well prepared Argonaut needed at least five white muslin shirts, five waistcoats, seven coats, 12 flannel shirts, 17 pairs of new heavy pantaloons and 18 new checked shirts. Upon arrival in San Francisco he discovered there was no way to transport, let alone keep, all these East Coast necessities at the diggings; he gave almost all of them way. Christman did keep one item–a journal of his adventures. In fact, many of the forty niners kept some sort of record of the gold rush. It is these reminiscences, notes and diaries that relay to us the texture and tone of the times.

The feelings and expectations of the California gold rush were captured in the music of the late 1840s. Stephen Foster was born on July 4, 1826 in Pittsburgh, Pennsylvania. He his best remembered for his songs of the romantic pre-Civil War Southern life. Oddly, he had no first hand knowledge of the South beyond one brief visit to New Orleans in 1852. The gold rushers knew him as the composer of "O Susannah," a minstrel song that the Argonauts made their anthem after changing a few words. Legend has it that a group of Salem, Massachusetts teenagers bound for San Francisco on December 23, 1848 first changed the verses of "O Susannah." One of the boys is said to have signaled his companions to begin singing their improvised verses to the crowd of well-wishers on shore. They began by singing a simple revised line:

▲ *Placerville in all its early glory, soon after its name was changed from Hangtown. The hodgepodge of hastily erected buildings and haphazardly planned streets gave gold rush towns their disorderly appearance.*

I came from Salem City with my Washbowl on
my knee...

As soon as the boys knew they had the crowd's amused and undivided attention, they continued the song with words written for the bon voyage gathering:

I'm going to California, the gold dust for to see.
It rained all night the day I left, the weather it
was dry.
The sun so hot I froze to death—Oh brothers,
don't you cry!
Oh! California, that's the land for me
I'm going to Sacramento with my washbowl on
my knee.

Soon afterward other versions of "O Susannah" were being sung by the gold rushers waiting to leave their homes or on the journey westward. Other contemporary songs had much the same sentiments. One went:

Then blow, ye breezes, blow!
We're off to Californi-o.
There's plenty of gold,
So I've been told,
On the banks of the Sacramento.

The songs were cheerful and optimistic, and those who left were unprepared for the adventures they would encounter during the trek to California, or the experiences of digging at the gold fields. For most of the Argonauts it was not a walk sun.

The 122 men of The Hartford Mining and Trading Company in Connecticut were everything New Englanders should be: orderly, sober, hard working and good citizens. They were also excellent craftsmen. Nothing that was wrought by man could not be either built or repaired by the assembled group of Yankee forty niners. On the journey aboard the ship *Henry Lee* the skills of the company were tested by Providence.

On March 11 the *Henry Lee* was three weeks into her voyage to California via Cape Horn when an unexpected storm struck in the late evening. None of the landsmen knew what to expect. Their sole knowledge of the sea came from books and tall tales, but they awoke to find out first hand what a sea born tempest was like. The New Englanders first reaction was panic as they tumbled out of their births and attempted to reach the hatchways in the wind tossed darkness. All round them there were terrified men yelling for help; some were just screaming while others were paralyzed with fear and silently praying. And then came a sound, a snapping, tearing, crashing so loud it was if the gates of hell had opened and were ready to swallow the ship whole.

In the purple dawn the men of the Hartford Company discovered the ship was still afloat. The storm did not damage the hull, only the masts. Masses of sea drenched canvas covered the deck, along with the tangled rigging and splintered wood. As fearful as the farmers, blacksmiths, machinists, shoemakers and harnessmakers were in the evening, they were as equally resolute in day; they now knew what had to be done. Tools were unpacked and work begun. In 12 days the masts were repaired and the *Henry Lee* was running before the wind, bound for a stopover in Rio de Janeiro.

When it is summer in the Northern Hemisphere, it is winter in the Southern Hemisphere. Not only did this mean winter conditions, which most of the Cape traveling Argonauts who set sail in the summer were unprepared for, it also meant incredibly stormy seas in the area around Cape Horn. These Southern tempests were as bad if not worse than the gale that struck the *Henry Lee*. Very few of the forty niners did any research on the conditions they would encounter on the way to San Francisco. They were taken by surprise at the 100 knot fierce frigid winds that tossed their ship about, like a bar of soap in a bathtub.

For the most part, the grueling aspects of the seaward journey to San Francisco were the demons of bad food, sea sickness and boredom. Food was a problem between stopovers because of the lack of refrigeration. It was a strange fare the Argonauts ate on the open seas. Filling the stomach was the first order of business, to this end new and barely palatable recipes were created. Lobscouse was a stringy paste made from salted meat, potatoes and hard bread. Hushamagrundy provided more fiber with its mixture of 70 percent turnips and parsnips. The rest of the concoction was made from ground codfish. Sometimes when dessert was served, it was a combination of hardtack cooked in molasses with raisins and cinnamon. When supplies ran low there was a three bean soup, with each portion consisting of mostly water, three beans and an inch cube of rusty pork. When the water grew foul, after being stored for several months in vats, it attracted all manner of

▶

Images such as this adventurous miner entering the wilderness on a handsome white horse to make his fortune may have accomplished as much as any professional modern advertising campaign by drawing young men to California.

insects, many of these bugs found their way into the soup. To kill the horrid taste, if not the creepy things floating in the water, vinegar and molasses were added and the potion was called switchel. No wonder there were so many cases of seasickness.

Boredom became a silent, unseen companion. There were only so many sunrises, sunsets and moonrises that any human can appreciate before they become as commonplace as the sky itself. When the passengers weren't betting on cards, checkers, poker, backgammon, what was being served for dinner or anything else to pass the time, they invented practical jokes to play on each other. After awhile even these amusements would fail to break the constant atmosphere of ennui.

On some ships the passengers attention, no longer diverted by natural marine wonders or artificial amusements, focused on the cause of their frustrations, the conditions aboard the ship and its captain. At the beginning of his journey Enos Christman had nothing but glowing phrase for Addison Palmer, captain of the schooner *Europe*. After six months at sea and with the ships provisions running dangerously low, Christman wrote in his journal that the captain had, "hardly capacity enough to take charge of a canal boat, let alone a vessel that plows the oceans. He is colleagued and identified with the mean and rascally owners." This gripe may have only been a difference of opinion between a passenger and the captain towards the end of a long voyage, but one month out of port and the Argonauts aboard the *Pacific* were contemplating mutiny.

As an inducement to get first class passengers bound for San Francisco to travel on the *Pacific*, Captain Tibbets promised they would eat the same food that he and his first mate ate. No sooner did the ship sail than Tibbets reneged on his deal and fed the first class travelers the same gruel as those who paid considerably less for the trip. Feeling that they had literally been taken for a ride, the passengers complained loudly. Tibbets countered the protest by threatening to set fire to the powder magazine and send the ship to oblivion. The protesters ceased their complaints in the face of an act which would bring an untimely end to all their endeavors. In order to keep his unruly passengers in line, Tibbets posted a notice declaring all troublemakers would be clamped in irons.

A month later mutiny became the topic of discussion among the Argonauts aboard the *Pacific*. It was sparked by the discovery of a cargo of cheese, butter, flour, pickles and other such delicacies hidden in unmarked containers in the hold. Tibbets told the passengers they would be fed the better provisions when he felt like it and not a moment before. Someone in the crew told the angry and ill fed travelers that the captain was going to sell the food in San Francisco at a tidy profit for himself.

A small secret meeting of the passenger's leaders was held and the discussion turned to their legal position if they seized the ship by force. Among the group of plotters was a thin and laconic man, one Mark Hopkins by name. Hopkins would go to California and decide that money could be made not in gold but hardware. His partner in the shop he would set up would be Collis Huntington, and together they would become half of the partnership that become known as the The Big Four of the Central Pacific Railroad. But now the nascent robber baron was a partner in an incipient mutiny. Hopkins quitely urged caution. A passenger named Dr. Stillman agreed, citing that the captain's narrow expectation of the passengers behavior was the same as he expected from his crew. "The master damns the mate, the mate damns the second mate, the second mate damns the sailors. Our captain, having passed his life among seamen, is incapable of treating passengers any other way." The group postponed the mutiny and decided to lodge a protest with the U.S. consul when they reached Rio.

As soon as they landed in Rio the irate passengers filed a complaint with the American diplomatic officers. Tibbets heard about this potential new affront to his authority, but instead of getting angry he held a conciliatory meeting with the protesters. This was a subterfuge. The captain was preparing to leave that evening after having secretly cleared customs an hour before the meeting. Upon hearing of this deception, the passengers descended upon the U.S. consul and obtained an order restricting the ship to port until the matter could be fully investigated. A few days later the American officials verified the Argonauts story and relieved Tibbets of his command. Weeks later the *Pacific* sailed with a new captain.

Not every ship had bad food or bad service. For example, the luxury ship *Edward Everett*, named after the president of Harvard, offered such delicacies as cheese, apple pie, and plum pudding. Elegant accommodations aboard the *Everett* also included a weekly paper, concerts, a board of health and a police department. But the *Everett* and the few ships like her were the exception and not the rule.

Those who chose the sea-land-sea route across the Isthmus of Panama had different set of problems to overcome. On a map it looked like a quick and easy trip by ship to Chagres, a short hop overland to Panama City and then an equally easy sail to San Francisco. What would really happen on this trip was anybody's guess. You paid your fare and took your chances. It was to become known as the "Gambler's Route," for the fare you eventually paid in money and hardship always varied from Argonaut to Argonaut.

In 1849 Panama was part of the Republic of New Granada, a neutral semi-autonomous state confederated in a union with Greater Columbia. Many of the treaties involving New Granada were negotiated in Bogatá by the Columbian president. In 1845 a convention was negotiated with the United States which recognized the right of free transit across the Isthmus for American citizens in return for formal recognition of New Granada. So the forty niners had no legal problems with this route.

Initially, the price for the trip to San Francisco, via the Isthmus, was about $300, but that turned out to be only the beginning. Crossing the Isthmus began at Chagres, a small town located at the mouth of the Chagres River and guarded by the ruins of San Lorenzo, a fort destroyed by Morgan the Pirate in 1670. The first 54 miles of the 75 mile trip to Panama City was not by land, but by water, up the Chagres River in dugout canoes called *bungos*. During the first few days of 1849 Argonauts paid the native New Granadans $10 for this privilege. After the first week of January the same *bungo* ride cost $40. They paid 50¢ for a whole chicken, cheap for gold rush cuisine, but they paid $2 for a pot to cook it in. These inflationary prices were to be expected wherever the gold hungry hoards went. They paid $2 for a hammock by the riverside and gave a blood tribute to the mosquitos and other creepy crawly night bandits.

Travelers progressed by river transportation from the mosquito and disease infested swamp town to the higher elevation and healthier country surrounding the towns of Gorgona in the dry season, or Cruses in the wet season. It was a seven day jungle journey filled with exotic wildlife and ridden with cholera, dysentery, malaria and yellow fever. For the most part the natives of mixed Indian, Spanish and Black parentage were amiable, treating their passengers in the same manner which their passengers treated them. Many an arrogant Argonaut discovered how unwise it was to anger his boatman; he could make a traveler's life unbelievably uncomfortable if not downright dangerous.

When the first mob of forty niners suddenly showed up at Cruses the natives of the sleepy little town, fled fearing war had been declared and the Americans were invading their town. For five days the local government pleaded and then negotiated with the inhabitants to bring their horses and mules out of hiding and rent them to the Argonauts. Once the natives were convinced of the Argonaut's peaceful intentions they rented their mules and horses to them. The travelers used these animals to carry their baggage to Panama City, where the last leg of the journey would take the forty niners by steamboat to San Francisco.

The Argonauts had every reason to believe there would be a boat waiting for them at Panama City. The U.S. Mail Steamship Co. had just begun postal service from New York to Chagres and the Pacific Mail Steamship Co. took the mail from Panama City to San Francisco. In between, the mail was transported via the same route across the Isthmus as used by the forty niners. Nothing stopped the mail from getting to San Francisco, but the gold rushers were not so lucky. Once the weary, hungry, bug bitten travelers reached Panama City a new set of ordeals were waiting to ensnare them.

At the time, Panama City was much less what one might have reasonably expected from its name. In the 17th century it was a major seat of Spanish power in the New World. In the middle of the 19th century it was a decaying backwater town. Thick jungle-like grass grew between the cobble stones that paved the streets. Abandoned stone buildings had deteriorated and fallen to ruin. Those dwellings still inhabited were in various states of disrepair. All of Panama City's faults would have been easily overlooked had a ship, any ship, been waiting in port ready to take the gold rushers on to the California valleys of golden opportunity.

In 1849 there were few ships and many passengers. The town was overrun with people and services were on the verge of breaking down. What hardships the forty niners endured on the trip across the Isthmus was due to being unfamiliar with the natural flow of life in a exotic land; the hardships endured at Panama City were due the familiar workings of man. Waits for space aboard a ship bound for San Francisco could be up to four months. The going price for a room in the local "luxury" hotel, when available, was $8 a week. As more travelers crowded the small city, those without a place to stay built temporary shanties that grew into small towns. The miserable little

In good weather and in the absence of danger the greatest peril to the ocean going traveler was boredom. The cures for this malaise were gambling and practical jokes.

one or two room shacks lacked rudimentary sanitation facilities. An odor would rise over the towns with the sun and remain in the air until it rained. In some ways they were similar to the hastily built towns in the gold rush country with one important exception, these men had nothing but time on their hands.

As with the Cape route finding something to do became a full time chore. Men went on hunting trips, attended cockfights, watched the local women, took plenty of sea water baths in the 90 degree heat and got in trouble. Drunkenness became the norm and fights were fought on a regular basis. One riot broke out when a group of drunken men entered a cathedral with their hats on, thereby offending the faithful church goers.

And then there were the illnesses. By the end of March there were 2,000 new inhabitants in Panama City. Sanitation went from barely passable to fatal. Many men fell ill. The strongest survived and the weak ones died waiting for the ships to arrive. The only way out was by ship to California or admitting failure before having even seen the gold fields and trekking home. For those who found this lush stopover more than they could bear, returning home empty handed was the only option. In spite of all these obstacles, most of the gold rushers traveled on.

The problem in the early days was the lack of ships for the number of travelers demanding transportation. Ships that plied the Pacific were built on the East Coast. The first steamship used as a mail transport on the West Coast was the 1,000 ton sidewheeler, *California*. It sailed empty from New York in October 1848 because the gold rush had not started yet. After a stormy passage around the tip of South America the *California* picked up 69 Peruvian gold rushers in Callao, Peru, sailing for an expected arrival in Panama City on January 4. Bad weather delayed the steamship's arrival until January 17 and waiting on the pier was an irate mob of Americans, demanding to be taken to San Francisco. There was room for 210 passengers and a third of the available room was already taken by the Peruvians.

Col. Mason's replacement, Maj. Gen. Persifor F. Smith, was in Panama City at that time seeking to travel to San Francisco with all deliberate speed. When Smith learned of "foreigners" taking up space that should rightfully go to Americans, he became incensed and an early leader of the "Americans only" notion which became prevalent in the gold fields of California. Smith fired off a letter of complaint to the U.S. Consul in New Granada. In it he promised to enforce the present and future laws of trespassing on public lands as they concerned persons who were not citizens of the United States. In reality, the letter was a threat to the Consul's secure job if he did not remove the Peruvians from the *California*. The ship's captain interceded with a compromise. The Peruvians would give up their steerage berths and sleep on deck, thereby providing accommodations for 250 Americans. She sailed on February 1 with 320 passengers, well over the legal limit. Almost a month later, when the *California* docked in San Francisco, her passengers became the first wave of forty niners to arrive at the gold fields.

This was the beginning of the trans-Isthmus route and a lot of bugs had to be worked out. Whalers that regularly plied the Pacific made the mistake of stopping at Panama City only once, and once was all that was needed. After being overrun with passengers and having the crew desert in San Francisco, many a ship's captain made a vow never to return to the gold rusher infested city in Central America. In the following years the situation improved until one week was the longest lay-over one had to expect in Panama City.

Noting the numbers of Argonauts who used the Chagres to Panama City route, in 1849 a group of businessmen obtained a charter from the government of New Grenada for the building and operation of a railroad traversing the Isthmus. Work was begun as soon as the the Panama Railroad Company was incorporated in 1850. But the inhospitable conditions in the swamps and jungles made construction slow and costly. When finally completed, in 1855, the trans-Isthmus journey took a mere three hours. It was a wonderment of the young industrial age. Many believe the Panama Railroad initiated premature notions of building the Panama Canal.

There were other sea-land routes that went through Mexico. These trails were favored by Southerners and especially by those who fought the Mexican War. The veterans claimed that Mexico offered low prices, plenty of supplies and easy passage. Language was no problem if the party took a veteran along because all veterans allegedly knew enough Spanish to get them through the travails one might encounter in that foreign land so close to home.

Some of what was carelessly stated was true. Much of the trans-Mexico route was easier than the all sea and trans-Isthmus routes. There were supplies available and in the beginning prices were cheap. Yet there was a certain wishful naivete on the part of the veterans. After just suffering a defeat and losing a large part of their country to the United States, many Mexicans were less than sanguine about having former enemies tracking across their land. Some Mexicans tripled the price of goods sought by the "gringos." Some who were not in the business of selling goods chose to prey on the gold rushers by taking up banditry. The Argonauts became targets of opportunity for the outlaws as the unsuspecting travelers marched through the mountain passes. As for the boasts of being able to speak the language, some of the former soldiers might as well have claimed to speak Chinese for all the help it brought them.

The two widely traveled Mexican routes both used Mazatlán as a terminus port. One route started by sailing to Vera Cruz from any American port. From there gold rushers went by horseback to Mexico City to restock on supplies. After a short rest they would continue their ride to Mazatlán, where they hired a small boat to take them to San Francisco. The other route entered Mexico at Laredo, Texas and proceeded by way of Monterey.

By the end of the year the estimated population of California had risen from 26,000 to approximately 115,000. Of the 89,000 immigrants, 41,000 had taken one of the sea routes and passed through San Francisco. Some of the 697 ships that carried them to the land of gold were abandoned by their crews who went gold seeking with the passengers. These ships all but rotted away in the bay until years later when they were refurbished to carry the gold rushers back home.

Over half of the forty niners took the land route because they were either too poor or too cautious to take chancey and expensive sea routes. When they traveled across the country it was in covered wagons called prairie schooners. These covered wagons were smaller versions of an ungainly Eastern freight carrier, the Conestoga. Even though they were constructed on a smaller scale, these newly built or modified farm wagons were still large wagons. Bows were looped over the box-like hardwood blue painted body. White canvas, frequently weatherproofed with a mixture of beeswax and linseed oil, was stretched over the bows. The red painted wheels were reinforced with iron rims. One does not get the patriotic effect of lines of red, white and blue wagons plying their way across the prairie from looking at old black and white photos of the period. To complete the portrait, the wagons often had either the company's name or words like "gold or California" painted with a bright color on the canvas cover.

The covered wagons were powered by animals. Mules and oxen, or sometimes a combination of both, were the favored animals for the long trip to the gold fields. Mules were to mountains as cars are to highways, and they could travel 40 miles per day on the flat country. The trouble with mules was their cost and the difficulty city folk had in handling them. Oxen were slower, only managing 15 miles a day, stronger, more reliable and would eat food that mules and horses refused. The animal engines of the prairie schooners were fueled on the gasoline of the open spaces, grass. And it was this fuel which kept the forty niners from venturing across the continent until the beginning of May; because before May the grass had not grown tall enough for the animals to dine.

At the western end of the railroads near Saint Joseph, Independence, Missouri and Council Bluffs, Iowa, makeshift towns mushroomed out of the prairie, filled with thousands of forty niners waiting for the grasses to reach the proper forage height, spring flooded rivers to go back to their normal height and the snow to melt in the winter blocked mountain passes. The frontier parasites sprouted in the temporary towns like weeds in a freshly fertilized lawn. The gamblers, con artists, prostitutes and thieves thrived until the time came for the gold seekers to begin their trek across the continent. When the Argonauts left the staging areas along the Missouri river, the towns disappeared like smoke in the wind.

The most widely used trail to the West ran from Council Bluffs, along the Platte and the Sweetwater Rivers to the valley of the South Pass, around the Great Salt Lake following the Humboldt River through the desert to the timbered slopes of the Sierra Nevada and finally into California. This is roughly the route followed by the tracks of the Union and Central Pacific when the transcontinental railroad was built in the sixties and seventies. Some 8 to 10,000 men went by one of the southern routes via El Paso, or one of the several trails that crossed northern Mexico. Whatever the route, the whole journey took about 100 days, if there were no unplanned interruptions.

At first it seemed it would be a three month long party. The gold rushers organized each wagon train along military lines, and each train might be composed of several companies with their own officers. The elected leader was called the captain, his subordinates consisted of one quartermaster and three or four lieutenants. All the major decisions: when to stop and start each day, where to camp and how to defend the camp were his responsibility, but in truth he had no legal authority. In the beginning, when things were going well, his lack of legality mattered little.

Little by little things came apart. The first to go was all the non-essentials. As the wagons got stuck in the prairie sand and mud, weight and later room on the wagons became a major concern and the surplus goods had to go. Chief on the list to be dumped was heavy mining and camping equipment, the stuff the Argonauts bought at the urging of crafty promoters. Closely following the useless contraptions littering the trail was all the extra clothing. It was thrown on to the trail by the trunk load. All a forty niner really needed was a pistol or rifle, food, elementary tools and three sets of clothing. Everything else was dumped to make room for fodder and water for the livestock. In some places the trail looked like a huge deserted outdoor flee market with pianos, harps, chests of drawers, tables and tons of fancy clothing scattered all over the countryside.

Indians were not a danger at this time. Although the wagon trains would form the traditional defensive circle each night, hardly any Indians attacked the steady stream of forty niners crossing their land. Occasionally a tribe might demand a fee for transit or once in while a some livestock was rustled in the dark of night, but the Indians of plains and prairies did not perceive these transients as a threat to their land and well being. That change in perception would come years later.

For those who started out from the Missouri River later, around June and July, food for their livestock became a problem. It wasn't that the grass had not grown, the problem was that the spring grass was eaten by the livestock from the earlier wagon trains. When grass was found on the prairie it took experience to know what to do with it, and among the throngs of bumbling amateurs experience was in short supply. To the inexperienced, clover, blue grass, herds grass and buffalo grass were all the same. But sheep, horses, mules and even bison have flexible snouts which allows them to graze almost to ground level. Oxen have stiff snouts and can only graze to an inch and half above the ground. Buffalo grass rarely reaches an inch and half. A herd of oxen can starve in a field of buffalo grass. The forty niners gained their experience the hard way, they earned it by trial and error. And they had a name for such a revelation, it was called "seeing the elephant."

The term "seeing the elephant" became firmly rooted in the American language of the 19th century during the California gold rush, but its origin was a decade earlier. Legend has it that a farmer who knew what an elephant was, but had never seen one, wanted to see the animal once in his lifetime. As luck and stories like these go, a circus had arrived in a nearby town complete with a herd of elephants. Off went the farmer to the nearby town to sell his eggs and vegetables and see an elephant. Just outside of town he met the circus parade lead by a pair of pachyderms. The farmer's joy in achieving his lifelong goal was broken by the panic instilled in his horses at the sight of these grey behemoths. They bolted, overturning the wagon, breaking the eggs and damaging the produce. When asked about this unfortunate turn of events the farmer's comment was, "I don't give a damn, I've seen the elephant."

Water was not everywhere on the prairie, and when found, it was sometimes bad. The Humboldt River was a subtle killer. As its water ran from the mountains it picked up and dissolved minerals. By the time the Argonauts reached the portion they would follow on the trail to California, it became successively brackish and alkaline. Any man or beast who drank its waters at this point would surely die. Like a bad comedy the worst was yet to come. Just when the water was so bad that it give off a stench that could drive mules away, the water vanished into the grey black earth and became the Humboldt Sink. After that the forty niners had 40 miles of desert to travel through before they reached the Carson River with its water and grasses. The reality of the Humboldt River and Sink was in direct contrast to the guidebooks published in the East that stated the Humboldt River was glistening with pure fresh water and "beautifully clothed with blue grass, herds grass, clover and other nutritious grasses." It was in the worst desert this side of hell, beyond the Humboldt Sink, that the majority of Argonauts saw the elephant.

In the spring and summer of 1849 cholera was the most dreaded of the forty niners' hardships on the trail and in some of the mining camps. It crept into the United States on December 1, 1848 aboard the packet *New York* which, coincidentally, was docked in New York harbor. Cholera was at epidemic proportions in Europe and the *New York* had sailed from London. A month or two later the disease showed up in the New Orleans ports. It was carried into the heartland by river boats to Saint Louis and from there to the Missouri River.

In reality cholera is not one disease but a term applied to a wide variety of short duration diarrheal diseases. In 1849 these outbreaks were due to contaminated food and unsanitary conditions found on the crowded trail. Campsites were abandoned without disposing of the garbage or infected clothing of the cholera victims. Flies, contaminated water and sometimes cholera patients themselves spread the bacilli to the next arrivals. Many believed they were fleeing cholera and going to the land of gold and health. In truth, they were bringing the disease with them.

The acute stage of uncontrollable diarrhea and vomiting lasts from 2 to 12 hours. In this time the victim becomes severely dehydrated, which in turn causes agonizing muscle cramps and a dry, blue skin. Many victims burst the blood vessels in their faces due to their convulsive retching. The next 24 hours determined if the sufferer either lived or died. If the patient recovered, the symptoms disappeared and he or she would be weak for months. Otherwise, the victim became increasingly debilitated and hardly had any noticeable pulse. Sometimes death took an unmerciful four days to arrive.

There are no figures on how many Argonauts died on their way to California from cholera, but the best estimate is about 5,000. Many of these were buried on the side of the trail and sometimes impatient companies would just leave them to die and infect or be buried by those who came later. The road along the Platte River became a graveyard, and on any given day in the spring and summer of 1849 a gold rusher would see at least two or three families burying their loved ones. Not everyone who took one of the three main routes to California had bad experiences. Some would arrive in California with no trials and tribulations to report, ready for whatever fortune had in store for them in the gold fields.

Whoever said getting there is half the fun should have been with the forty niners. Not only was there fun to be had getting there but then there was the fun of gold mining to contend with. Before they left for California, and after months of arduous travel, many of the Argonauts still held to the notion that gold prospecting was about as difficult as picking apples or digging clams. Some even gave away coins to their friends who declined to take part in this once in a life opportunity saying, "There'll be plenty more where I'm going." The reality of the situation was about to dawn on each newcomer to the gold sites.

THE DELUGE

In 1849 chaos ruled the gold country and improvisation was the rule. Everyone came to make money, one way or another, no one had time to think of building permanent structures. Build-

The mining areas of the American West and British Columbia.

ings, government and law enforcement were all quickly constructed with no thought of permanence, since most of the Argonauts did not expect to be in California more than a year or until they had "made their pile."

In the small mining towns with names like: Whiskey Diggings, You Bet, Poverty Bar, Long Bar and other extemporaneous designations, buildings were erected out of any cheap material available. The most common structure had a green lumber frame with canvas used for the walls and roof. Some "buildings" were really hastily converted covered wagons. Once in while a log cabin or two were constructed by families who still maintained the pioneer tradition of the East. Wood frame structures were more common in the large towns like Sonora and Placerville. Several of these structures were actually produced by sections in the East and shipped around the Horn. In the later 50s a few stores and financial buildings were made from brick or stone with iron shutters on the windows. This form of building became a necessity where fire was a frequent occurrence. In some communities, buildings were burned to the ground as fast as they could be built.

There was no city planning. The layout of the town resembled the jumble of streets in medieval towns. In some camps everything was built on the smallest amount of land possible, leading to small crooked streets with scarcely room for a man and a horse to walk abreast. Other camps spread like modern suburbs over two or three miles. In the gold country these camps grew helter-skelter, like wooden flowers planted by a deranged gardener.

Col. Mason was lucky. On May 1 he left the governorship and California without a care or worry. What passed for government in the last year was a fragile arrangement at best. In the newly acquired land there was only a military government superimposed on the existing local government, which in turn was based on a loose governmental arrangement with Mexico. No one in the new government planned for the tidal wave of Argonauts that swamped all the poorly staffed and underfinanced services. The flood of new arrivals had to work out their own solutions.

Mining claims were the most pressing issue in the gold rush country. Each camp and small town would hold a meeting and elect officers who in turn would draft rules which all those assembled would vote upon. The law was a hybrid of settlers' rules for land and livestock, European mining codes, Mexican mining ordinances and Wisconsin lead mining regulations. In other words, anyone who had any experience with land and mineral claims contributed to the local mining claim law.

The law of 1849 progressed from a claimant being anyone who cleared the topsoil off any portion of the bar of a river, as was the custom in 1848, to each camp determining the size of the claim according to richness of the local diggings. In a gold rich site the area could be as little as 100 feet, in poorer areas it was larger. Each miner was allowed only one claim and he had to prove his right to the claim by leaving his tools on the site and work it a set number of days a month, with time off allowed for selling the gold at some distant town or city. Enforcement of the camp's claim rules was up to a either a presiding judge, a committee or a mass jury consisting of any members of the camp who wanted to attend at the time.

Crime, a growing concern after 1848, was handled by a camp jury of 12 or more men. If the case was popular enough the whole camp became the jury. There was always a judge and prosecuting and defending attorneys, but the criminal law was the law of the district, which varied according to location. Punishment for the guilty was limited due to the lack of jails. Depending on the offense, a person was either whipped, branded, banished or executed. Some may think that banishment is not such a bad punishment, but in a gold rich camp it meant loss of one's claim and starting over in some area where news of the offender's crime was not known. To many a criminal, banishment meant the end of his search for gold in California. From that point on it was a choice between returning home or finding another venture. For some, this meant the continuation of the criminal life.

The gold rush country extended in the North from the town of Lassens' Ranch on the Sacramento River near Deer Creek to Mariposa south of the Merced River. Sonoma was on the western edge, while the eastern boundary was marked by an imaginary north/south line running through the western shores of Lake Tahoe. In 1849 the heart of the mining country was called The Mother Lode. Here is where the Argonauts came in search of fortune.

Newcomers to the improvised land were unprepared for the disorganized state of affairs that greeted them. Communications between mining camps were difficult. Not only were there no roads, but the trails ran over around and through canyons, foothills, ravines and mountains. Distribution of supplies, such as dry goods, tools and food started out like many of the gold rushers in San Francisco and traveled by river transportation to Sacramento, Stockton and Marysville. From these three distribution points goods flowed into larger mining towns like Placerville, and from the larger towns the supplies seeped into the rough country and hundreds of smaller camps. Situated wherever there was gold, these small camps were sometimes hidden at the bottom of chasms. The only way to find them was through direct knowledge of their existence or by practically falling into them by accident. When supplies finally reached one of these towns they tended to be very expensive.

Communications were further frustrated by the polyglot of languages spoken by the forty niners. By June California's population had reached 30,000. In April, May and June about 11,000 gold rushers had passed through San Francisco and of that number about 80 percent were not American citizens. By the year's end 25 percent of the 89,000 forty niners had come from foreign lands. They hailed from Mexico, Chile, Australia,

China, Hawaii, (not part of the United States at that time) Panama, Peru, France, England, and Tahiti. Even if you ran into someone who spoke English, it was likely that you might not understand what he said because of his accent, be it regional American, British or Australian.

Communications between traveling companions broke down as soon as they reached the golden land. Many of the companies which survived the trip over land and sea disintegrated once they reached the beginnings of the gold country. Like ideas about life, everyone had a different idea about gold seeking. Some wanted to rush to the diggings as soon as possible, others wanted to make sure they had enough supplies, some wanted to go the Southern gold country and others wanted to go North, some decided to become shopkeepers and some overwhelmed by the strangeness of it all wanted to get drunk before they attempted anything.

Later in 1849 the notion of cooperative association caught on with those who had placer mined the streams the year before. As the newcomers arrived looking for streams to pan the "old timers" formed small companies of three or four partners to divert the water running through a set of contiguous claims and expose the bed for easy and dry access. Later groups of veterans formed associations and began crude vein or lode mining ventures. In 1855 the Californians would call this form of hard rock mining "quartz mining," by then it would be a capital intensive field demanding a great deal of technical knowledge. The division of labor was becoming specialized and this would spell the beginning of the end for the freelance or solitary miner.

Stories of spectacular gold finds were lacing the swirl of confusion that met the forty niners who arrived in California, and it was these mythic sized tales that dispelled the gold seeker's early discouragement. The most famous tale of the California gold rush has been told and retold so many times that it ended up as comedy bit in the musical *Paint Your Wagon*. It seems a long winded preacher was saying quite a few words over the grave site of a deceased miner, and as expected the small gathering became fidgety. Some of the kneeling mourners squirmed in place while others reached down, grabbed some freshly dug earth and absentmindedly sifted the soil through their fingers. Suddenly, one of the congregation closely examined the dirt in his hands and yelled "colors!" The preacher promptly dismissed the gathering, the body was hastily put off to one side and everyone began digging in the grave. Needless to say, they found a large quantity of gold. And then there was the story of the Southerner and his slave. They both had the same dream about finding gold under a particular cabin in the camp. Both of them bought the cabin and dug up the floor; it yielded almost $20,000 in gold. Then there was the report of the little girl who found a seven pound gold nugget, or how about the miner who shot a bear off a cliff and when he went to retrieve the body also found a quartz ledge richly laced with gold. Almost everyone had heard of a story about some fabulous gold find.

Christman, the well prepared Argonaut who had given most of his belongings away in San Francisco, heard stories of incredible gold strikes before he left for the diggings. At the beginning of his 45 mile trek to the Southern diggings, Christman and his companions met several disheveled miners heading toward San Francisco. They told the eager Argonauts a tale of woe about mining and not being able to make a go of it. But the newly arrived group was not deterred by these hapless fellows; after all, the gold rushers thought, they were probably unlucky. There was no doubt that Christman and his two friends were on their way to discovering their fair share of the gold.

The Christman party finally arrived at the camp near the Merced River weary and hungry, with aching muscles and blisters on their feet. A slow realization of what a miner's life would be like dawned upon the Argonauts when they saw the extortionate food prices. Christman had started out from San Francisco, after selling all his non-essential clothing and tools, with $85. Now he had $27. If Christman had to pay the camp prices for food he would soon be broke in less than two weeks. He and his companions had to find gold in a hurry. On the following day the group split up and Christman teamed up with his friend, McGowan. On the first day they panned a grand total of about $1.50 in gold dust. The next day yielded about $1.25. It was easy to see they would starve at this rate, so the group banded together again and moved 30 miles south to the gold mines near Mariposa.

Things were changing in the gold country. The pan was giving way to the rocker and the rocker was giving way to long tom and sluice. Gold was harder to find; larger quantities of debris and gravel had to be processed in order to make what one man and pan could find the year before. Unless you were extremely lucky in finding and working your claim, it was no longer a question of simple placer mining. This is not to say the gold was running out, in fact, the total output for the year has been estimated at 2,500,000 ounces,

▲
Although oxen would eat forage that mules and horses could not, the large strong beasts would starve on the prairies when the grass grew less than an inch and half tall. Many gold rushers discovered this fact the hard way, by first hand experience.

much more than 1848. But in human terms the daily average gold production per man was a half an ounce. There were too many people and the easy finds were in inaccessible locations.

Unable to afford a new rocker or even the materials to construct one, Christman and McGowan built a makeshift rocker out of scrap wood and a perforated iron sheet. Two years later Christman would give up gold hunting, stay in California and go back to his old job working as a printer with the added responsibility of serving as a deputy recorder for a county government. For many, the search for gold did not pan out. With the new awareness that there was a limited amount of gold to be found in California came a wave of xenophobia, which in turn unleashed outbreaks of racial and ethnic discrimination that haunted California for years to come.

Because of their native cultures and tongues the French, Mexicans, Spanish Americans and Chinese tended to stick with their own kind. Around the world people fear those who do not mingle like "ordinary folk," and over time this fear turns to hatred. The first groups to feel this animosity were Mexicans and Spanish Americans. It was readily noticeable to the forty niners that the Latin Americans were much more skilled at mining than the average gold rusher; their higher level of expertise was due to the experience gained in their home countries. In the Northern mines, because of their small numbers, the Latin Americans were simply driven off their claims, in the Southern mines they banded together for protection and even formed towns like Sonora.

The French gold rushers were a different matter. Although the average American forty niner knew almost nothing about the European continent, its people and its customs, the Frenchmen embodied all things hated about Europe.

In December 1850 Louis Napoleon, then president of France, announced a lottery that would send the poor to California to seek their fortunes. In reality, he was getting rid of possible enemies by sending them out of the country and at the same time lining his pockets with the proceeds of the lottery. Many of the women chosen to go to California were prostitutes, who gave the impression to the naive Americans that all French woman were of loose morals.

While a few of the Australians that came to California were convicts, there was a wide spread belief that all the gold rushers from down-under were criminals. During the last part of the 18th century, the British settled Australia by sending English felons there in order to relieve the overcrowding of British jails. Australia was so distant from any other civilized place that escape was difficult. The island continent also had abundant natural resources and rich soil capable of sustaining a large number of people. By the late 1840s, however, the colonial administration and the Australian legislative councils had passed a resolution refusing to allow convicts to be sent to any part of the colony. They were sent to Tasaminia and Norfolk Island instead. Yet in California the belief persisted that all Australians were by their land of origin, criminals.

The Chinese, on the other hand, with their alien Asian culture, were truly despised by the latter-day gold rushers. Prior to 1852 most Americans in California were cordial to the Asian gold rushers, but in that year more than 20,000 Chinese flooded the California gold fields. When the Chinese gold rushers landed in San Francisco there was a support system waiting for them, organized by the Chinese merchants who had emigrated earlier. These organizations provided the new arrivals with supplies and transportation to the diggings. Some of the Chinese would perform contract work in the gold fields for these organizations which led to the the mistaken belief that the Chinese workers were slaves.

The Chinese gold rushers had a knack for working abandoned placer sites and making them pay off. It was their hard working, simple living and thrifty ways that set the Chinese apart from the profligate Americans. The Chinese spent little of their gold in the towns, thereby avoiding the high prices and returning home with much of their hard earned fortune intact. And then there was opium. Where the American miners would get drunk often and raise hell, the Chinese gold rusher would smoke his opium, lie down and dream. It was too foreign, too strange and too weird, for some of the Americans. At first their jealousy took the form of playing practical jokes on the Asian gold rushers. The Chinese wore their hair in pigtails. Cutting these pigtails off to make Americans out of the Chinese was considered very funny in the gold rush country. But the jokes soon took a more hateful turn with incidents of seizing claims, unfair taxes, violence and murder occurring with some regularity.

In 1850 the Foreign Miners' Tax of $20 a month was passed. This tax was chiefly aimed at

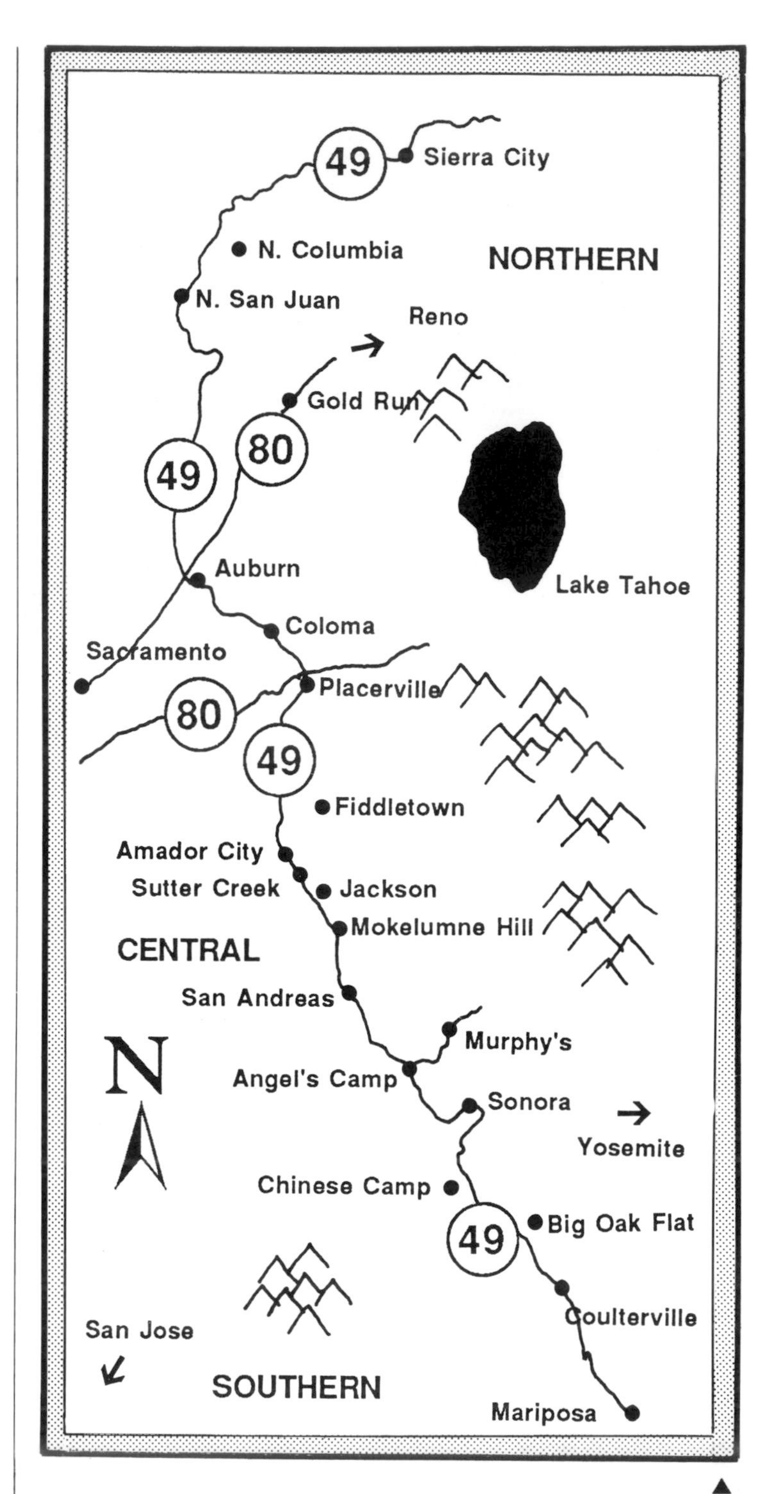

California's "Mother Lode": Route 49 was named in honor of the "forty-niner."

the Chinese and Mexicans, but it affected all the other foreign miners as well. By 1853, due to the protests of the English and Irish miners, it was lowered to $4 month. The Chinese were very patient and although some left California to return to China many remained, unassimilated, after the placer gold ran out. They became a permanent part of shaping the future of California.

Those who fared the worst were not foreigners, for they were truly native to the land, but the Indians. Prior to the discovery of gold in California

▲
Many Argonauts found that two or more men could mine more gravel than one, so they formed small companies to work jointly held claims.

the Mexicans, Indians and frontiersmen all lived in a pastoral peaceful coexistence. In the first year of the gold rush the Indians were welcome as workers; in the next year the Indians were driven away from the gold fields; by late 1850 the Indians were driven into the high Sierras and near starvation conditions. These were not warrior Indians. They had been raised in paradise. What followed was a nightmare of children being kidnapped, women forced into prostitution or sometimes the victims of gang rapes, men enslaved by alcohol and the occasional massacre. Unlike the foreign gold rushers, the Indians had no land overseas to return to. This was their land and their tale became a tragedy of paradise lost.

So lasting was this legacy of prejudice that 80 years later a popular book on the gold rush stated, "France sent several thousand lying men and corrupt women at public expense. Australia obliged with a few boatloads of ex-convicts. Mexico provided its quota of gamblers..." And so it goes...

All this intolerance did not make any more gold appear at the mines nor did it really increase the per miner wealth. However, it did breed distrust of all newcomers. A large number of the Argonauts were running away from some circumstance in the East. Their bad habits and anti-social ways were given free reign in the gold rush country, that is until they were caught and subject to miner's justice. This situation inspired a popular rhyme of the time:

Oh, What was your name in the states?
Was Jeffers, Johnston or Bates?
Did you murder your wife and flee for your life?
Say, What was your name in the states?

CHANGE AND GROWTH

Founded by Captain Townsend, leader of a group of Wisconsin prospectors, the town of Rough and Ready started as a campsite near a valley stream where the Argonauts discovered a profitable placer deposit. Townsend had served under "old Rough and Ready" Zachary Taylor during the Mexican War and he had the canvas on every wagon painted with the slogan "Rough and Ready." The name of the town was a natural choice since all the early makeshift buildings had the name already painted on their canvas walls.

Townsend built himself a more permanent

structure out of logs. It was, in fact, a large cabin that served as his home as well as the courthouse, saloon and town casino. When court was in session the judge had to caution the dealers to keep their voices down so the case being tried could be heard.

In April 1850 there was a town meeting called to discuss the new federal taxes. One miner was so incensed and so eloquent he convinced the whole town to secede from the Union. In the next few months the town elected a president and a secretary of state. When July 4 rolled around the nationalist fervor of the new city state dissolved with the celebration of the Declaration of Independence. The citizens of Rough and Ready were ready to be Americans again. It was only years later, after the secession movement was but a memory, that the federal government learned of the short lived independent state.

Throughout the fifties the town boomed and its population reached about 6,000. Then, in 1859, the placer gold ran out. At the same time there was a fire and most of the 300 buildings burned to the ground in the conflagration, leaving few buildings to remind future generations of its heyday. Other towns such as Sonora, Placerville and Jackson prospered, declined and survived into the 20th century to become real communities. However, a small town called San Francisco, fed by the gold rush, fitfully grew into something quite fabulous. A quiet change was occurring in the small port during the early hurly-burly days of 1849. While much of the population was transient, in town a short while before going off to the gold fields, some of the natives and new arrivals made the decision to stay and provide equipment, supplies and capital to the gold rushers.

Prices for goods were determined by scarcity. It was true that merchants were charging for whatever the market would hold, yet this was due in some part to the prices that the ships' captains were charging for goods they delivered to San Francisco. The captains would hold auctions on the beach for the goods they had transported to the gold area. Their main concern was to quickly sell the goods and leave as soon as possible. One reason for this haste was the fear that other ships waiting to get into the port might be carrying the same articles, and the first to sell would get the highest price. Another reason was the possibility that the crew would desert the longer the ship was anchored in the gold rush country.

Often, because of the lag time of at least three months between the time an order was placed for goods and the time the goods were delivered to San Francisco, the price would have fallen because all the other merchants ordered the same item at the same time. As a consequence, merchants in San Francisco experienced huge swings between scarcity and over-supply for many items. As a result, prices would fluctuate wildly. Unlike Eastern businesses, there was no coordination among the merchants to keep supply and pricing at an even level. It was everybody's intention to make money from the gold strikes, one way or another. There was no time for notions of organization.

San Francisco was a flea bitten town. In 1849 the greatest number of permanent residents consisted of fleas. They were everywhere, but mostly on the forty niners who would unwillingly take them to the gold fields. San Francisco was also a canvas town. Stone was not used for construction in the early days due to the shortage of labor. Wood was very expensive, when it was available, costing $1 for a 12 inch board. Often, abandoned ships were scavenged for their wood, with whole decks torn up and reassembled as parts of different buildings. But there was a great deal of canvas replacing the wood wall or missing roof. The fleas loved to live in the canvas, lying in wait for the next human meal to come by.

Brannan, the former Mormon, made money from this state of affairs. He didn't rush off and pan gold because he saw other opportunities. His first order of the day was the opening up of a general store near the mill. It was stocked with provisions that the miners would need, commodities such as food, picks, shovels and iron pans. The last item was used, of course, in panning, and Brannan had bought up all the available pans at the beginning of the San Francisco rush for 20¢ each and sold them from anywhere between $8 to $16 in gold. In the same vein, Brannan cornered the market in carpet tacks when San Francisco was mostly built of wood and canvas. For a short time everyone who constructed a shelter in the growing town used Sam's carpet tacks and he made a small fortune with his foresight.

The medium of exchange in San Francisco was gold since the amount of coins in circulation was very small. Depending on supply, the value of gold went from a low of $6 to a high near $18. The median dollar value was about $9. This fluctuation in rate explains much of the free spending habits of the Argonauts. John Henry Brown, San Francisco's inn keeper commented on this phenomenon: "Most people had an idea that gold dust would depreciate in value, judging by the quantity of which was brought into the city; consequently, I would pay out the gold dust as fast as possible, fearing I might lose by keeping it." The price of gold finally stabilized in 1854 when the United States mint opened a branch in San Francisco. Thereafter the rate of exchange for gold was fixed at $16 an ounce.

One investment which did not depreciate was real estate. At the beginning of the gold rush a fair sized lot might be bought for $17 or a few barrels of whiskey. A year later that same lot would sell for $45,000. Rents followed the inflationary spiral of San Franciscan real estate. Corner buildings were going for $6,000 a month with one room offices renting at $1,000. One of the prime instigators in the rise of rental rates was the gambling operations, which could afford to pay incredible prices for choice locations. They offered the miners the same chance to make a fortune as the gold fields, almost none. But the gambling operations, as other suppliers, were making a small fortune themselves.

With all this money changing hands the public treasury was broke. In August 1849 there were no policemen, jails, public offices, hospitals or graveyards. That month John W. Geary took over the office of alcalde of San Francisco and commented to the gathered townspeople: "Public improvements are unknown in San Francisco. In short, you are without a single requisite necessary for the promotion of prosperity, for the protection of property or for the maintenance of order."

Geary, true to his analysis of the growing city's problem, began his tenure by raising money for the empty treasury. His most successful scheme was the imposition of a large license fee on the city's gambling operations. With money flowing into the treasury, funds became available for a municipal court, a police force and the purchase of a ship to be used as a city jail. These three new services were badly needed, for civilized conditions in San Francisco had deteriorated to a point where no one believed it was possible to establish law and order.

Gangs with names like the Hounds or the Sydney Ducks had free reign of the city. At the core of the Sydney Ducks was a close knit group of former convicts, deported from England to Australia, who managed to leave the colony to participate in the California gold rush. After a short time in the gold fields the convicts discovered that mining was hard work and they returned to their former occupations. This core group was joined by others with the same criminal inclinations. Most notable of the newer members was a contingent of demobilized soldiers from New York. Many of these men were products of New York City's Tammany Hall, the last word in 19th century urban corruption.

The new forces of law and order did not eradicate the gangs. They did, however, manage to occasionally crimp their style. Yet this crimping of the criminals' style was a mere band aid. There were six major fires in the wood and canvas city from December 1849 to June 1851. During these fires gangs like the Sydney Ducks would loot the burning buildings. Honest San Franciscans believed that the gangs had created their looting opportunities by setting the fires themselves. On the evening of May 3, 1851 a fire broke out and quickly spread through the business district. Because of its speed many believed it was yet again the work of Sydney Duck. This was a bad fire. By morning it had destroyed a district that was three quarters of a mile long and a quarter of mile wide. Estimates of the number of buildings involved in the conflagration varied between 1,500 and 2,000. Darting among the flames members of the Sydney Ducks were seen looting. Several were killed in the act by outraged bystanders. It was now time for the citizens to rise up and let their will be known.

It was new, it was different, it was brutal and it seemed to get the job done. On June 9, 1851 approximately 200 citizens, including many business and professional leaders, formed an association called the the Committee of Vigilance. Its members were known as the Vigilantes. Their first test came on the night of June 10 when a former Australian convict named John Jenkins was caught red handed lowering a safe, which he had just removed from someone else's store, into his rowboat. His captors were two members of the newly formed Committee of Vigilance. Jenkins was swiftly brought to the committee's headquarters. Within a half an hour he was tried and found guilty by a jury of 80 Vigilantes. The irrepressible Sam Brannan, who was one of the committee's leaders, ran out of the building and announced the news to the crowd which had gathered while the "trial" was in progress. Jenkins was quickly taken the Custom House and hanged from a beam until he died.

After three months of operations the Vigilantes had hanged four men, whipped six more, forced scores of gang members to flee the city and even handed 15 criminals over to the police. By their very actions the Vigilantes underscored the corruption and weakness of those charged with maintaining the peace. Around the United States feelings were mixed over the use of vigilantes to bring about law and order. Most of the Western newspapers backed the Vigilante Movement, while in the East and North such actions were viewed with disdain with the feeling that San Francisco was a safer, if not completely crime free city, the committee ceased to function, although it never formally disbanded.

Short term memory did not function well in San Francisco during the gold rush. Five years later the Vigilantes reactivated the committee in response to the city's return to lawlessness. Over 800 killings had taken place since the last meeting of the Committee of Vigilance. Then, in 1855, a federal marshall, Gen. Richardson, was murdered by a gambler named Charles Cora. It was a petty affair. Richardson's wife and Cora's mistress, Belle, had a minor argument at the theater. This altercation led Richardson to pass a few sneering remarks about Belle's character. Which in turn led Cora to vow to get even. A few days later he did, with a single shot. Cora was arrested and brought to trial. Cora claimed to have righted a grievous wrong and half the jury, along with the judge, believed him. This resulted in a hung jury and mistrial. Cora was set free. The next public murder blended this homicide and political corruption in one outrageous incident.

James King, former banker turned crusading publisher and editor of the *Evening Bulletin*, exposed crime and corruption without a thought to his own personal safety. The police and the courts constantly came under the *Bulletin's* scrutiny. According to King the police failed to arrest criminals and the court failed to prosecute many of the alleged felons. He was not alone in his disgust of official corruption, but he was the leading voice calling for reform.

Richardson's murder brought a violent denouncement of Cora in the *Bulletin*. At the same time King attacked an editor of a rival newspaper, the *Sunday Times*. James P. Casey learned his political ethics in New York City's Tammany Hall. He put this knowledge to good use by rigging his election to the town council. King learned how Casey stuffed the ballot box and published this accusation in the *Bulletin*, with the additional allegation that Casey was an ex-convict who served time in Sing Sing. Casey was furious. He demanded, and did not receive, a retraction of these charges. Instead, King threatened to publish documentary evidence to support his claims. Casey's furious reaction was swift and sure. He shot King in broad daylight, in front of witnesses, on May 14, 1856. He drifted in out of a coma for four days before he finally died.

Casey was arrested and an angry mob formed outside the jail demanding that the killer be turned over to them. The sheriff appeared and assured the doubtful crowd that Casey would be given a prompt and impartial trial. Coincidentally Cora was residing in the same jail awaiting his retrial for Richardson's murder. Instead of letting justice take its normally convoluted course, a new Committee of Vigilance was formed. This time the committee had nearly 10,000 members. An around the clock guard of several dozen men was immediately set up to prevent either or both men from leaving the jail.

The standoff continued for two days until Gov. J. Neely Johnson arrived from Sacramento, the new state capital. Johnson wanted to take Casey and Cora to Fort Gunnybags for trial. The Vigilantes refused to let this transfer of prisoners take place. They told the governor that the city government was controlled by Casey's friends and a fair trial would not happen under the current conditions. It was reported in some circles that Johnson told William T. Coleman, veteran of the 1851 Vigilante Movement, "Well go to it old boy." The governor also urged haste since he was under a lot of pressure to resolve the situation.

▲
Nothing that had ever happened before prepared the sleepy little town for the hoards of Argonauts who descended upon it in 1849.

The next morning 24 companies of Vigilantes, complete with cannon, gathered in front of the jail and demanded the two infamous prisoners. In the face of the massive show of force the sheriff reluctantly surrendered Casey and Cora. They were taken by carriage to the committee's headquarters to stand trial. After three days of testimony, and upon hearing of King's death, the Committee of Vigilance found both men guilty. The next day two events happened simultaneously: King was buried after an elaborate funeral and Casey and Cora were hanged in front of a crowd of over 10,000. As was the case with the 1851 Committee of Vigilance, the 56 Vigilantes disbanded after a couple of more hangings and quite a few banishments.

In spite of all the crime San Francisco in the middle 1850s was a lively city of 50,000. It was seen by the rest of the country the same way New York City is thought of today, exciting but dangerous. It had many theaters where plays, musical extravaganzas, operas, ballets and minstrel shows were performed. There were also race tracks, saloons, amusement parks, gambling houses and houses of ill repute. And most important of all, it had women.

In 1850 a little more than 7 percent of California's population was women. In the gold rush country the percentage of females was down to 2 percent. In the late 40s most of the women came from Mexico, Chile and Peru. They were the first wave of a group of women who have been described as "fair but frail," "neither wives, widows, nor maids," the "soiled doves" or to paraphrase the words of Mark Twain, "they were virtuous according to their lights although their lights were somewhat dim." Some were out and out prostitutes, others saloon companions and still others the companions of miners as long as they were treated well.

Those who came from Mexico settled mostly in the Southern diggings. A few making the trip further up to San Francisco. The women who came from Chile and Peru were the most exploited of these female immigrants. They were penniless and could not afford the crudest of boat transportation, yet they were allowed passage aboard some rather elegant ships at the captain's discretion. The captains knew the women's fares would be paid for in San Francisco by the propri-

▲
The first test of the new Committee of Vigilance's resolve came with the apprehension and hanging of John Jenkins.

etors of fandango houses, who would pay anywhere from $100 to $300 per woman. Fandango houses were also called dance cellars in San Francisco because they were mostly located in the basement of buildings. Inside the fandango house the recently bought women would dance the fandango, a Spanish-American dance done in three quarter time. Most of the miners only watched the women dance while waiting for more personal and lusty entertainment. Most of the women working in the fandango halls smoked cigars, a habit the American miners found unsettling.

Later in the gold rush women from France and China began to appear in California. The French women were much more stylish than Latin American counterparts and their jobs were more glamorous. They worked in gambling houses, sitting with lonely miners for an ounce of gold or sleeping with him for a pound. Many of these women were sponsored by Louis Napoleon's lottery that sent the poor to California to seek their fortunes. The Chinese women were for the most part slaves. They were unwillingly brought to San Francisco and made to ply their trade in small rooms called cribs.

Ah Toy was the one of the few exceptions. She paid her own way to San Francisco. Once there she started her own prostitution business. She was a one woman bordello. Some of the Chinese men felt Ah Toy was too much competition for the women they had brought to this country, so they conspired to have her deported by paying a man in Hong Kong to claim he was her husband. At her deportation the judge asked Ah Toy why she came to San Francisco. She proudly replied, "...to better her condition" Feeling this was a reasonable reply for any immigrant, the judge allowed her to stay. Despite her occupation, Ah Toy believed in justice. Several times she brought charges against her clients for cheating her by paying in brass filings instead of gold dust.

Then there was the fabulous Lola Montez, born Eliza Gilbert in Limerick, Ireland. She came to California in 1853 and entertained the woman-starved miners at the camps. The men loved her, but not because she was a great performer, for she wasn't. She was, however, a legendary woman who had left Emperor Ludwig I of Bavaria and several other lovers and husbands behind in Europe. In a land where women were rare, she shone and sparkled like gold.

Not all the women who migrated to California were of "loose morals." Only the majority of the first group to arrive. It was a romantic time and men who tired of expensive one nighters sought the companionship of a wife. All a woman had to

▲
Lola Montez was notorious, she a woman and she had no talent, but none of that mattered in the gold rush mining camps where there was less than one woman for every 10 men.

▶
Marie, Susie and Nellie were three Chinese prostitutes who posed for this rare 1866 photograph. The ratio of Chinese men to women at the diggings was over 12 to 1, and only 10 percent of the women were not prostitutes. It was said in China that respectable women did not come to America.

▲ *Lunch time at a long tom operation. The few women who made the trip with their husbands had to find work doing laundry, cooking meals and delivering snacks. The man with the outstretched hand is either indicating that the woman is his wife or that he is very hungry.*

▶ *Unlike other placer mining techniques, hydraulic mining was an American first. It was invented during the California gold rush and remained basically unchanged throughout the 19th century. It consisted of building up enough pressure by funnelling a stream of water through a series of flumes. The resultant pressurized stream of water was sprayed against a hillside where it melted the hill away much like an ocean wave erodes the wall of a sand castle. Needless to say, the debris was rocked and panned for gold.*

do was show up in a mining camp to be showered with proposals and gold. If a woman's husband died, it was rare for her to remain a widow long. Many are the tales of women who married the chief mourner at their deceased husband's funeral the next day.

Like today's personal columns, notices appeared in the newspapers seeking husbands. Since it was a sellers' market, these notices sounded more like contracts than enticing advertisements. A woman seeking a husband would list the chores she was willing to perform, the extras she learned, reading and writing, for example, an idea of the age range for the prospective husband and how much gold she expected to be given before she married.

Some women crossed the prairie with their families, endured terrible hardships and, in some cases, watched their children and husbands die. When they arrived in gold fields they set up house and sometimes helped by rocking the gold rocker. Then there were those women who found the wild nature of gold mines a nurturing climate for their latent entrepreneurial skills. Luzena Stanley Wilson was such a woman.

Dirty, with a month of trail dust clinging to them, the Wilsons arrived in Nevada City, California in the summer of 1849. After hours of

washing the dust from her face and children's skin, she looked around and realized there was money to made in Nevada City, and it wasn't in mining gold. Her husband quickly built a crude shelter while Luzena drove stakes for the legs of a large table into the ground. The next day she had 20 men paying $1 for dinner. Soon they had a hotel called the "El Dorado" that served from 75 to 200 boarders a week. All this money began collecting under the floorboards of the Wilson bedroom. At one point there was over $200,000 just laying there doing nothing. Luzena put the money to work by lending it out to the miners at 10 percent interest a month. She and her husband might have become one of the richest families in America if a fire had not swept through Nevada City leaving them and 8,000 miners penniless.

What could be called the "salad years" of the California gold rush lasted from 1848 to 1852. A miner might have made a small fortune with only a pick, shovel and pan. The key words are might have, for many miners spent what they made on food, drink and women. After 1852 gold mining changed. The sites that were easy to mine were running out.

Placer gold production had reached its peak in 1852. At the same time teams of miners and private companies were digging tunnels to reach prehistoric placer deposits hidden under sedimentary hills composed of a combination of dirt, lava and rock. Tunnel mining was costly and speculative. Some tunnels were dug over 1,500 feet long without finding enough gold to make the enterprise profitable. A new, less expensive labor intensive technique was needed.

Hydraulic mining originated during the California gold rush and is one of the notable American achievements in mining techniques. It is still used today. All the other forms of mining used in California were either borrowed from other countries and other times or reinvented through ignorance by the forty niners. Essentially hydraulic mining, or as it was known in the gold fields, hydraulicking, was nothing more than aiming a jet of pressurized water at a hill suspected of covering an ancient placer deposit and washing the covering mound of debris away. To accomplish the high water pressure needed, miners first built large flumes on top of a tall adjacent hill to carry water from a diverted a nearby stream to the bottom of the hill to be removed. As the water cascaded down the flume, the trough narrowed. At the bottom, the pressurized water coursed through iron pipes to a nozzle where it shot out to dissolve the hill. The loosened ancient placer debris was caught by a series

▲ *A placer miner's paradise, this picture shows all forms of California mining in the 1850s. Left to right: One man operating a rocker; a lone Argonaut panning a stream; a lode mine in the side of a hill with a loaded car being pushed out of a tunnel; a bucket lifted out of a shaft; new tunnels being dug into a hillside; debris shoveled into a long tom; gravel shoveled into a sluice system.*

of large sluices where the gold was separated. This could be risky business and occasionally miners who accidently walked in front of the jet of water were killed.

The gold rush was becoming less egalitarian and more capital intensive. Equipment had to be sent from Eastern cities, making price out of

reach of the ordinary miner. Earning a large amount of money from the fruit of one's labor in the gold fields was being replaced with a hired work force. The cost per ton of dirt handled was going down while the yield was rising. Hydraulic mining was a much more complex operation than the informal companies of forty niners that were formed to dam streams and rivers and operate small sluices.

Concurrent with the rise of hydraulic mining came the traditional form of tunnel gold mining called quartz mining. Quartz mining also required a heavy capital investment and an organized labor force. Stamping mills were also needed to to crush the mined gold ore and prepare it for more complex gold recovery processes. The whole operation demanded capital investment, a high level of technical expertise and years of work on a particular vein before the mine was profitable. Not the kind of industrial dream originally envisioned when the gold rush began. A new phase had begun, and for all practical purposes the gold rush of individuals had come to an end.

Of those who started the California gold rush Marshall's luck ran true, all bad. He sank deeper into despair when he should have taken an active part in the gold rush. When he did act, his handling of the situation was inept. In an incident

▲ *Early in the California gold rush the need arose to process gold bearing rock mined from lode deposits. Many of the Argonauts attempted to use elegant looking, complicated machines designed in the East and created by inventors who had no idea of the task involved. The simple solution for small-scale operations turned out to be the Mexican and South American arrastra. The gold ore was placed in the track while the mule dragged the stone and crushed the ore.*

involving his ineffectual protest over the vigilante slaughter of Indians accused of murdering several miners from Oregon, Marshall incurred the enmity of Indian-hating whites who recently moved into the town of Coloma. It wasn't that his stand was wrong, just ill thought out and executed. Thereafter, none of the miners respected his posted claims around the mill, although his claim to the land had a solid foundation in law and local custom. Marshall refused to chase the claim jumpers off his stake. Instead, he sank into a drunken stupor of self pity after various ill fated attempts to recreate his first gold find by the mill.

After large and powerful mining interests prevented Marshall from recovering his lost properties, he toured the West giving rambling lectures about his discovery and selling autographed cards. The monies he earned were enough to keep him from starving and drunk most of the time. In 1872 he was granted a $200 a month pension by California legislature, which was halved in the following session and cut off altogether in 1878 when he appeared drunk in the state assembly asking for an increase of his stipend. He died in 1885 a destitute and broken man and buried on a hill overlooking the sawmill. A 10 foot statue of Marshall holding a nugget was erected in his honor. It cost a mere $25,000.

Sutter faired slightly better than his sawmill partner. He saw his dream of an agricultural empire evaporate with each worker going off to the gold fields. He did nothing to stop this migration. Produce went unharvested and rotted in the fields. His livestock strayed through broken fences and were eaten by hungry, failed miners. Some believe he should have tried to lure the disenchanted gold rushers into his employment by paying top dollar. Instead, he drank and ran off to find some gold for himself. Sutter found that his two successive gold expeditions were completely

▲
The gold rush was almost over in California when the newer, sleeker, faster ships began running service to San Francisco.

unsuccessful. He was a bad manager and an equally bad role model. After finding any substantial quantity of gold his workers would go off on drinking binges. If the boss allowed himself to get drunk, the workers reasoned, so should they. Sutter returned to his fort as a failed miner only to find a long line of creditors waiting for him. In a clumsy attempt to retain his holdings, he turned some of his largest properties, near the developing town of Sacramento, over to his son, August. His son was not cut from the same "creditors be damned" cloth and sold much of the holdings to pay the debts. This caused an intense row between father and son which ended when August had a family friend travel back to Europe and collect the remaining Sutter family.

Eventually Sutter saw his lands overrun and much of his property stolen. In 1858 the U.S. Supreme Court held that much of the Mexican land grant was invalid. This ruling forced Sutter into bankruptcy. He settled in Pennsylvania with his wife who came to share in his fortune and ended up sharing his misery. He died in Washington, D.C. on June 18 after failing to get Congress to act in granting him a stipend.

Christman returned home in 1852 after mining a little more than 100 ounces of gold. It was enough to pay off his debts and marry the woman he corresponded with for three long years. He was not a wealthy man, but considered himself to be a man rich in experience and maturity.

By the end of 1849 California had grown from 14,000, not counting Native Americans, to almost 100,000. By 1860 that number had increased to 380,000. Although other gold rush areas in North America would gain in population none had such a large jump as the California rush. Over 250,000 people went to California to seek their fortunes in the early stages of the gold rush. A few made great sums of money mining gold. It has been estimated that close to $400 million was mined in the first seven years of the gold rush. A few like George Hearst, father of William Randolf Hearst, began their fortunes by investing in mining operations. The start of his fortune began by learning the ins and outs of mining in the California gold fields. Later, he would use his knowledge to invest a mere $450 in a silver mining operation and make a fortune. And a few made a windfall by becoming merchants and businessmen in a land of golden opportunity.

In terms of gold recovered and money made, the California gold rush was not the richest strike in North America. It was, however, the greatest gold rush in terms of human drama and the historical development of the United States. The slow westward movement in the 1840s was accelerated to a frenetic pace by the gold rush. California became a state, and in order to link the state with the rest of the nation the transcontinental railroad was finished in 1869. Although only some of the gold rushers made a fortune, while others lost their health or their life, it was a once in a lifetime adventure, and they would not have missed it for the world.

▲
Fortunes were made and empires built in the raucous days of the California Gold Rush. Wells Fargo started out as the Western branch of American Express. They came to California in 1852 and challenged the Adams & Company express service for the business of safely transporting gold and parcels to and from the gold sites and out of state. By 1855 the competition ended when Adams & Co. was forced into bankruptcy, leaving Wells Fargo with about 90 percent of the express business.

CHAPTER 6

NORTHERN BOOM

By 1855 California was inundated with gold rushers, old timers as they were now called, with no place to mine. As the gold rich exposed placer deposits dwindled, capital intensive industrial mining techniques took over the business of gold discovery and recovery. The pan and rocker miners could not compete with the hydraulic and quartz mining companies. Yet as old timers squeezed the last bit of gold from the remaining placer deposits, they dreamed of participating in another fabulous gold rush.

THE RIVER

Suddenly news reached San Francisco that gold was discovered on the Kern River, over 200 miles from the nearest mining camp, at the extreme southern end of the Sierra Nevada Mountains. Immediately, 5,000 men bolted to the new gold site. It was all in vain. The reports were not even exaggerations of some small nugget of truth. It was all a hoax. Three years later new rumors about a gold find somewhere near the "the North Pole" began to circulate in California. Although the geographic location was exaggerated, the discovery of gold was not.

British Columbia is the westernmost province of Canada. Its coastal mountains represent the greatest mountain mass in Canada. Even though the province is north of Washington state, the coastal climate is often mild and humid in the winter and dry and warm in the summer. Like California, the exact shape of the region was not known until the late 18th century.

At first it was thought by European cartographers that British Columbia contained the western terminus of the mythical Northwest Passage, a navigable approach that cut through North America and would connect Europe with Asia. Many Europeans thought of the Northwest Passage like a river as wide as a sea stretching from the Atlantic to the Pacific Oceans. Sir Francis Drake is reputed to have sailed along the coast of British Columbia in 1579 and found no such geographical feature. Upon his return to England Drake stated, "There is no passage at all through these Northern coasts, which is most likely, or if there be, that it would be unnavigable." But myths die hard and it would take two centuries before the book could be closed on the notion of the Northwest Passage.

The first recorded exploration of British Columbia was a brief landing on the coast by the Spaniard Juan Pérez in 1774. Two years later Captain James Cook, on his third and last great voyage, mapped the general outline of Northwest America, which included British Columbia, and landed at Nootka on Vancouver Island. He noted in his log that after an exploration of the coast, the Northwest Passage did not exist. While waiting for some minor repairs to his ship to be completed, he traded miscellaneous items with the Indians for a few sea otter pelts.

On his return trip to England Cook stopped in Macao on the southern coast of China and discovered that sea otter skins were a tremendously profitable item. Unfortunately, Capt. Cook's life was brutally ended in the Sandwich Islands when he was stabbed in the back and hacked to death by the natives. It takes a lot to kill a legend. But his narrative log returned to England where it was published and caused quite a stir among the sea faring nations. In the following year a veritable fleet of trading ships appeared off the coast of British Columbia. The British and the Spanish argued over the rights to the fur trade. The English, already embroiled in a war with 13 rebellious colonies on the East Coast of North America, were not about to lose a dispute over the northwest coast of the continent, especially not to the Spanish. And so British Columbia became a British territory occupied by fur trappers and ruled by the North West Company, which later merged into the Hudson Bay Company.

The demarcation of British Columbia's southern boundary was settled by the Oregon boundary settlement of 1846. It was one of President Polk's four great achievements during his term in office. Before 1846 both Britain and the United States had joint occupancy of the territory. But popular sentiment in America, summarized in the slogan, "Fifty Four Forty or Fight," gave the issue a political importance that it was lacking in Great Britain. As with all successful negotiations, there ensued a crucial compromise which fixed the international boundary along the 49th parallel rather than the fifty four forty latitude further north. Fort Victoria, the far western headquarters of the Hudson Bay Company, was relocated north of its former site near Vancouver, Washington to the southeastern tip of Vancouver Island, British Columbia. Partly as a bulwark against further American expansion, Vancouver Island was made a British colony in 1849. It was thought highly unlikely that the Americans would have any interest in British Columbia. After all, they were in the throes of a bad case of gold fever in California and all British Columbia had to offer was farm land, furs and cattle.

In 1857 Victoria was considered to be the most typically English town outside of England. Now it was the capital of the British possessions on the Pacific Coast. Surrounding Victoria were rich farms and large country estates. Men would pass their evenings slowly smoking their pipes and trading stories told to them by sailors whose ships had recently visited raucous San Francisco. It was comforting to know that such madness would never touch this northern British paradise. Near Victoria, Indians were adopting the white man's tools and skills. This led to an amazing burst of creative energy in the form of intricately carved and brightly painted totem poles. Little did they suspect their whole world was tottering on the precipice of change and about to take an irreversible plunge. This plunge would occur on the incredible Fraser River; a river that demands the tribute of immense respect from all who stand on its cliffs.

The Fraser is born of glacial ooze and dissolving snow high in the Canadian Rockies. When seen on a map, all 850 miles give the appearance of a very large lazy S. Before the Fraser makes a turn to the south at the northern elbow of the S shape, it is a peaceful, green, ever widening river. After the Fraser turns to the south, running parallel to the Coast Range, it picks up speed as it is joined by many small tributaries. Now the river becomes occasionally compressed by narrow gorges formed by steep cliff. This constriction causes fierce rapids to appear. Three quarters along its body, before the final elbow of the S, the Fraser is joined by another large river, the Thompson. The combined water flow makes the river swell a and few miles later it takes the final turn toward the Coast Range and the Pacific Ocean. It is here where the Fraser must use its brute strength to pass through the mountains, some which are taller the Rockies. It does so by slicing a narrow channel into the rock, thereby forming a long, black, roaring canyon. When the first explorers discovered the river they laid flat on the ground and peered over the edge of the canyon, lest the dizzying height and roaring power of the river hypnotize them and entice the men to walk off the top of the gorge into the angry waters below. Once the Fraser passes the Coast Range, it again becomes peaceful as it widens out into the coast plain. The Fraser is one rough river.

McDaniels and Adams were two tough prospectors, only McDaniels was meaner. Together they had discovered a large quantity of gold dust on both the Fraser and Thompson rivers. This was not the first gold to be found in British Columbia. Smaller finds were reportedly found on Vancouver Island and in other locations by agents of the Hudson Bay Company. The site was not by any means a small find, but it was not a large gold strike either. By California circa 1848 standards it was average, but that was big enough to kill for, and that is exactly what McDaniels did to poor Adams. After taking possession of Adams' share of the gold dust his first stop was an old trading post at Olympia, in the Washington Territory. He spent his bloody gold dust as if it was just the tip of a huge gold mine. It must be an enormous claim the townspeople thought, for someone, a stranger, to waste his gold in this out of the way, run-down trading post. Alcohol did McDaniels in. Little by little people heard the whole story of the find and Adams' murder. What happened to McDaniels afterward is unknown.

Rumors of the Fraser River gold spread to Oregon, and from there into the California gold rush country. Many of the California miners had no idea that the British had any possessions on the West Coast, not that this fact would stop them from finding any way possible to get to British Columbia. It was ship time at Panama City all over again, except now there was enough passenger space to go around if you had the money.

There is another explanation for the news of the Fraser River find traveling to California. According to an alternative report, it was the shipment by the Hudson Bay Company of 800 ounces of gold to the U.S. Mint in San Francisco that triggered the gold rush and not the homicidal, spendthrift and loose lipped habits of McDaniels. There is good possibility that both events occurred concurrently, which would account for the large and spontaneous wave of gold rushers traveling to British Columbia. About 25,000 miners from California, Washington, Oregon, Latin America and Hawaii rushed to mine the Northern gold.

As with the early days of the 1848 gold rush the initial reports were exaggerated, and as in 1848 there was gold to be found. Some of the early miners who arrived from Washington and Oregon found a good amount of gold. At one site three miners rocked $500 in six hours and at another a group of 100 miners mined $25 a day, per man. These turned out to be the exception, not the rule. When these stories where heard in San Francisco they were told as gospel. A miner who had an unhealthy case of gold fever was interviewed by *Harper's Magazine* on his way to British Columbia. He became an innocent forty niner all over again. "Where else in the world could the riverbeds, creeks and canyons be lined with gold?" he said, wanting desparately to believe, "Where else could the honest miner 'pan out' $100 per day every day in the year?" He had it bad. At the same time the old forty niner standard had taken a new twist:

Oh I'm going to Caledonia-Thats the place for me;
I'm going to the Fraser River, with a washbowl on my knee.

British Columbia was known as Caledonia among the gold rushers in California. Even those miners who were used to tall tales about hidden placer deposits in out of the way places believed the tall British Columbian tales. Deep in their hearts they knew there was more gold to be found and the Fraser River was a good enough place as any. The poor town of Victoria never saw such a crowd.

It was exactly what James Douglas, governor of Vancouver Island and regional chief factor of the Hudson Bay Company, was afraid of, thousands of men swamping his town and creating the same kind of freewheeling, anarchical situation which swept through San Francisco and the California gold rush country. It was going to be very disordered and very un-British.

As his name implies, Douglas was a Scotsman, but he was born in Demerara, British Guiana on August 15, 1803. He attended school in Lanark, Scotland and began his working career in Canada, at the age of 17, with the North West Company. When the North West Company merged with the Hudson Bay Company, in 1821, Douglas became a fur trader. He rose from clerk to chief trader in 1834 and later became chief factor in 1839. In 1845 he became senior member of the board, managing the company's operations west of the Rockies. In 1851 Douglas became governor of Vancouver Island and for awhile he held both governmental and commercial posts simultaneously. Although he was always an active man, Douglas was now going the have the busiest year of his life.

As early as December 28, 1957 Douglas asserted the British government's right to all gold deposits in British Columbia by proclaiming several licensing regulations: all miners had to obtain permission to work a claim and that permission cost £10 or $50 a month, all vessels had to obtain a license to enter the bays and creeks of British Columbia, with the fee varying upon the size of the vessel, finally, there were import duties on all goods brought to the mainland. As a money making scheme these regulations were highly successful; by the fall of 1858 the Hudson Bay Company's safe almost burst trying to contain $2,000,000 of American money. There was one little snag that Douglas had overlooked. In order to obtain the necessary licenses and pay the fees, the gold rushers had to first stop at Victoria.

By the spring of 1858 the little town of Victo-

ria had 10,000 gold rushers camped outside its stockade waiting for passage to the Fraser gold mines. By the summer, this number would swell to 20,000. It was not at all what Douglas expected. At first he thought these men were the garbage that had been forced to leave San Francisco by the Vigilantes. In his report to London, Douglas wrote: "They are represented as being with some exceptions, a specimen of the worst of the population of San Francisco; the very dregs, in fact, of society." To some extent this was true; there were contingents of alcoholics, gamblers, thieves, pickpockets, prostitutes, con-men and ex-cons. Yet the behavior of the American gold rushers was milder than Douglas had expected from listening to the the tales that drifted in with the ships and sailors who had been in California. Douglas ended his letter by stating: "Their conduct here would have led me to form a very different conclusion."

▲
Gold rushers had to be rugged men in order to prospect in the hard and beautiful Fraser River country.

Nevertheless it was an invasion, or at least that's what Douglas thought. The only way to deal with the threat of anarchy on Vancouver Island and on the mainland was to be resolute in the imposition of order. He was going to do things the British way, which to the American way of thinking was a bit stiff. In a confrontation with a spokesman representing 100 miners who refused to take the oath of allegiance that accompanied the filing of claims in British Columbia, Douglas showed how tough he could be. A dozen or so unruly miners had gathered in front of his estate mansion. Douglas proceeded out of his house to meet the miners and took a firm stance on the question of the oath. Defiantly the miner asked, "Suppose we came and squatted?" "You would be turned off," answered Douglas, showing no emotion. Thinking the governor's poker face was a sign he was bluffing the miner continued, "But if several hundred came prepared to resist, what would you do?" "We should cut them to mincemeat, Mr. ________, we should cut them to mincemeat." In truth Douglas had already overstepped his authority. His official jurisdiction did not extend to the mainland. The British on Vancouver Island knew this, but nobody was going to tell the Americans, and this minor illegality was not going to stop Douglas from maintaining order.

The Indians watched the white men encroach on their lands, eat their salmon and harvest their gold, the very gold they had recently learned to mine. They retaliated by stealing tools, food, clothes and of course gold from the miners. No matter how many sentries were on guard at night, something would always be missing from the mining camp in the morning. No doubt events would have stayed at this low level of pilferage had news of an Indian war in Washington not reached the tribes on the Fraser.

British Columbia's Indian/American troubles began with an ambush of a 400 pack horse caravan near the Fraser canyon. Three men out of 160 were killed and close to 100 horses were stolen. This success began a sniping spree by the Indians. One by one the gold rusher's bodies floated down the the Fraser. The Americans immediately did what they knew best, they formed Vigilante committees. Their second act was to ban the sale of liquor, often the cause of trouble, to the Indians. The third act was to go get the troublemakers.

A company of 40 men left the town of Yale to get revenge for the death of two Frenchmen, who were found floating down the river. A short distance up river, at the smaller crossing point town of Spuzzum, the two groups clashed. The Indians lost seven men and were routed. The Americans burned the Indian's wooden ancestral tombs to the ground. Each of these tombs was an elaborately carved chest of drawers, about 15 feet long and close to 3 feet high. The tombs were laid out on the site like the village of the dead. The craftsmanship was said to have surpassed that of the Indian totem poles, with some tombs dating back over 100 years. With this act it looked like the Americans had started a bloody Indian war.

A group of 160 miners quickly formed another Vigilante company and besieged the Hudson Bay Company's store in Yale. They demanded rifles from the proprietor who, following company policy, always accommodated the Indians. For Hudson Bay it was a matter of good business, the Indians supplied the company with skins; the very animal skins that brought a great profit in Asia. In the face of an angry mob of Americans, however, he went against policy; the proprietor could not refuse their rather strong request.

The two Vigilante groups joined forces at the burned out burial site at Spuzzum and now over 200 men advanced up the river looking for Indians. At the mining site of China Bar the Vigilantes found five wounded men hiding in a cave. They were part of much larger party of miners who were attacked and had fled down river, suffering losses from ambushes every few miles. Sadly, the injured survivors told their rescuers everyone else was killed. The angry mob surged up river to find the culprits. There are two variations of what happened next: the first states there was a firefight which resulted in a stand off, the second states the Indians, realizing they were out numbered, raised a white flag of surrender. In either case the result was the same; a truce was declared until Gov. Douglas could show up to sort out the mess.

There were crises brewing on the mainland and Douglas was prepared to deal with them. A contingent of royal engineers had recently arrived in Victoria. They were sent to the northwest ostensibly to survey the new international border, but the governor planned to use the musket wielding engineers as a show of British strength in the mining camps. Douglas commandeered a stern wheeler and loaded it with some of his own troops, the royal engineers and a cannon. He sailed up the Fraser to Yale, which was the first stop for the armed dog and pony show.

Amazing but true, all the combatants were holding to the truce when Douglas arrived. To demonstrate the power of the office governor, he promptly unloaded the troops and cannon in a very impressive military manner. The governor initiated a peace conference and both Indians and Americans participated. The parley resulted in a

▶ *There wasn't much to the early town of Yale. It was only a place to sleep, buy provisions and unwind on a Saturday night.*

continuance of the prohibition of liquor sales to the Indians, a cessation of Indian attacks on the miners and in a show of allegiance to his new office and the Crown, ordered the company store to sell goods to the miners and Indians alike at reduced prices. For anyone found guilty of any crime less than murder the penalty was loss of all claims in British Columbia. This left the convicted man's site open for other miners to freely, and without penalty, claim jump. To insure the continuance of law and order at the mine sites, Douglas appointed several justices of the peace.

Gold on the Fraser was damned difficult to mine. Because of the treacherous Fraser canyon and river, panning and rocking was anything but easy. Unlike most rivers which subsided after the spring floods, the Fraser would rise again due to the melting of the high mountain snows. Mining could not commence until early August, when the river's waters were at their summer low. This left a window of several months for gold mining until late October and the arrival of winter, with its freezing cold and deep snows making mining dangerous if not outright impossible. Some of the more impatient gold rushers returned to California in July rather than wait for the river's floods to subside. After October thousands more left British Columbia finding the work too hard and mining fields too regulated.

At its height, the gold rush mining operations on the Fraser accounted for no more than 10,000 men spread out over a 200 mile area. The next year gold mining on the Fraser became more technical, following the pattern of evolution seen in California. With their sluices, flumes and wing dams, the mechanized mining efforts of the gold rushers paid off by yielding three times more gold than in 1859. This was accomplished by 3,000 or so miners who stayed through the winter at Victoria. Prices for goods took the same course as in California and skyrocketed. It was difficult for many of the marginal gold rushers to hold on to their hard earned gold. At the same time Douglas lowered the mining fees to £1 ($5 a month) or one large yearly payment of £5 ($25). For the miners who stayed on this was a help.

The story of the Fraser River gold rush was partly a clash of cultures. The Americans were suspiciously viewed as expansionist after the recent Oregon settlement. The British had a very structured way of governing and doing business. America, in less than a hundred years, had already developed a freewheeling culture, where societal boundaries were based on opportunities and not class system. The gold rushers who went to British Columbia viewed the licences and fees as "English fogyism," and the British were seen as extremely lacking in American enterprise when it came to taking advantage of their gold resources. From the British perspective the Americans were undisciplined, unruly and eager to annex all valuable land at a moments notice. These are oversimplifications of some fairly complex emotions and issues, nonetheless, they are good generalizations of the motivations that were involved. But the British Columbia gold rush was not over yet. There was one more chapter to be written further north.

THE ROAD

While working the gold on the Fraser River, some of the miners wondered where all the placer deposits originated from. There was a growing suspicion that somewhere up river, north along the body of the S shape, there had to be some form of mother lode. The miners first stopped at the Quesnel River, about 300 miles north of Yale. As with many gold finds the first on the scene mined the most. Between late 1859 and 1860 nearly 1,000 men were making in the range of $10 to $60 a day. Some of the first nuggets found weighed as much as 8 ounces.

Many of the Indians that inhabited the Quesnel River region knew very little about placer gold mining. They would come singularly or in small groups to watch the Americans panning for gold. One particular Indian, a son of a tribal chief, was totally fascinated by the process. He sat and watched Peter Dunlevy pan gold for about an hour, then he moved closer and sat and watched. After another hour had passed he struck up a conversation with Dunlevy. The chief's son drew a map on the sandy ground and told the miner he knew a place where there was much more gold. With that he placed a dot on the map to point out the spot. Not knowing the area Dunlevy asked for more precise directions to the gold site. Lacking any common frames of reference between them, the Indian agreed to show the miner and a few of his friends the location if Dunlevy would meet him in 16 days. At the appointed time the two men met and the chief's son took Dunlevy and four of his friends to Horsefly Creek, which quickly became the site of the richly productive Horsefly Mines. But there was one nagging question: If following the Fraser this far north could yield more gold mines, what if they followed the river closer to its source?

Many of the miners were functionally literate. They could read and write their own names, maybe a bit more than that, but they were not scholars. Spelling was not their strong suit in the bridge game of life. So when they came to a region known for its vast herds of North American reindeer, the Cariboo, they spelled it Cariboo. John Rose and Sandy McDonald were two such men.

In the fall of 1860 they trudged up river, past the North Fork of the Quesnel, to the headwaters of the Bear River. The land these two prospectors ventured into resembled a hurricane tossed sea, with jagged peaks and steep valleys, suddenly frozen at the height of the storm. Softening the sharp and hard landscape were green bands of growth, trees, bushes, lichens with creeks and streams tracing lacy patterns through the valleys. On a sunny summer's day the air was clear-blue and crisp-bright. It was just the kind of place where one might stop at a creek and notice brilliant bits of sun yellow nuggets concealed among the pebbles. And that's exactly what Rose and McDonald did; they went from stream to stream finding larger and larger quantities of gold dust until they arrived at Antler Creek. Here, all Rose and McDonald had to do is look into the clear water and find tiny gold nuggets winking back at them in the late afternoon sun. That first day Antler Creek yielded an ounce of gold; the second day it yielded four more. The next day snow interrupted the mining.

There was no way to keep this gold strike a secret. By early March close to 1,000 miners had abandoned their Quesnel claims and dug themselves earthen caves by the shores of Antler Creek to wait out the snows and the beginning of warmer weather to start of the mining season.

Again the news traveled quickly. Newspapers in England and America carried exaggerated accounts of gold found in incredible amounts and again the rush was on. Gov. Douglas realized that even though these cycles of British Columbian gold rushes might not last forever, there were going to be permanent settlers all along the Fraser and in the remote Cariboo region. In 1861, communications and transportation from Yale to Cariboo meant river transport when weather and river flooding permitted, and pack animals carefully plodding along narrow canyon trails. This was fine for the rugged placer miners, but soon there were going to be farmers. Lumber industries would eventually spring up and exploit the forests for lumber. Douglas foresaw the need for government, roads and public expenditures in what was essentially a wilderness. A road was needed to further develop the area after the gold rushes were over.

In the spring of 1861, about a month and half after the first shots of the American Civil War were fired at Fort Sumter, Douglas toured the land where he thought the road should be built. He came to realize that even though the future site of the road, by the banks of the Fraser, would start out as an easy job, once he past Spuzzum the canyon side grades were steep and dangerous. Yet he knew where there was no ?purchase? on the canyon walls the road builders would blast the rock away to make them, where it was impossible to adhere at all to the canyon the road would cross to the other side of the river and where there was danger of avalanche the road would detour inland. No matter how much the cost, the road would be built.

Once more Douglas exceeded his authority. His dispatch to London in October of 1861 was not a letter asking permission to build the road but notice that it was going to be built. He started his communication by telling the Duke of Newcastle that he was going to: "...push on rapidly with the formation of roads during the coming winter." His intent was not only to extend trade and commerce into the interior but also: "...defeating all attempts at competition from Oregon." This competition was not likely considering the United States was now engaged in a civil war. But this fear of American power extending itself into BC was a bit of an obsession with Douglas.

To insure that his road project proceeded without delay the governor closed his dispatch by simply stating: "I have in these circumstances come to the resolution of meeting the contingency and raising the necessary funds by effecting a loan of twenty pounds in this country. In taking this decided step, I feel I am assuming an unusual degree of responsibility; but I trust the urgency of the case will justify the means." None of these unilateral actions should have surprised anyone in the British government, after all wasn't it his take charge attitude that got him the commission as governor in the first place? While Douglas was busy raising money for his road into the interior, another big gold strike was discovered. It was to become the most famous bonanza of them all.

William Deitz and Ed Stout went where no white men had gone before, to the head waters of the Willow River. While they were prospecting they came upon a small creek which they named William's Creek after Deitz. It looked like the same old story, another poor mining site in the picturesque wild. There was, of course, a few gold nuggets and dust panned and rocked from the loose gravel laying on the top of the creek bed; but it was definitely nothing to write home about.

▲ *During the early days of the Fraser River gold rush there were all types of pack animals and trains used. Here, a team of oxen are preparing for a return trip to their coastal base.*

Nevertheless, the two partners stayed with several other miners panning gold and hoping for a big strike. By now William's Creek was becoming known as Humbug Creek among the other miners in the area.

There is a cloud of doubt cast over who did what at William's Creek in the following weeks. Some historians believe it was Deitz and Stout who made the fabulous discovery. Others say it was a couple of miners that stayed to pan gold alongside the two partners. In any event, what happened next was a case of having learned a thing or two from the California gold rush. Someone remembered that the reason hydraulic mining was invented was to get at the gold hidden under tons of debris. The buried gold was there because of the ancient river bed under all the dirt and rock. Well, there weren't too many places for an ancient river to be hidden in this rough, stony country except under the hard blue clay of William's Creek. So someone decided to dig and dig and dig. They dug non-stop for almost two days, without any clue of what they might find. Finally they hit bottom. It is said by some that Deitz and Stout emerged from their shaft with over $1,000 in gold. This was the beginning of the Cariboo gold rush in 1862.

Between 1862 and 1864 over 200,000 ounces of gold were mined from the Cariboo region. The majority of that gold production came from William's Creek. By 1864 the banks of the creek were honeycombed with shafts reaching down into

the blue clay and wet, slimy tunnels running helter-skelter underground and across the ancient gold rich stream bed. The legendary Cariboo Cameron dug a shaft 80 feet down and came out a millionaire. He employed 80 miners, paying them an average of $14 a day. About 4,000 men worked William's Creek during this gold rush, and the number would have been far higher had it not remained difficult to reach and expensive to live there.

While work continued on Douglas' road, now named the Cariboo Road, there were some unique schemes to better transportation into the region. In the beginning of the gold rush most of the freight was carried by pack horses and mules across the rough terrain. Later, mules were favored for never having to be roped together or kept under reign, their sure footing on narrow canyon trails, their ability to carry more weight than horses and because they could follow a trail with very little guidance. Some operators thought they were pretty damn smart. Mule packers organized their animals into trains of 16 to 40 mules and could average about 10 miles a day, making the run from Yale to Quesnel in a month. As in the California gold rush, those who supplied services, such as the mule train operators and packers, made more money than most miners.

Frank Laumeister thought he had a better idea. He wanted a pack animal that would be especially useful in the dry season, need less water and carry more weight. His solution was U.S. Army surplus camels. In the spring of 1862 he bought 21 camels and transported them up the river by boat. Soon everyone was talking about these strange beasts that could carry a half a ton of goods and move a quick 30 to 40 miles a day on the Cariboo trails. Laumeister was seeing easy money dance before his eyes. He should have used his nose instead.

There is no other way to put it... camels stink. Usually their owners and handlers get used to the odd stench, but to everyone else they reek. If it was only a matter of people being upset with the odor, then it would have only been a major inconvenience. However, the horses and mules were driven crazy by the stench. A tiny trace of camel smell in the air was enough to send the other pack animals fleeing into the woods and flying off of canyon cliffs in an effort to escape the nauseating smell. Needless to say, where dreamy dollars bill signs once floated above Laumeister's head, lawsuits now descended upon him like an avalanche. Laumeister managed to sell a few of the camels and the rest he let go on the Thompson Flats. They sired no offspring and the last camel died in 1905.

Work was hard in the Cariboo region, yet somehow the miners found time for entertainment. None of the great female stars of the day visited the miners. There wasn't a married woman in the entire Cariboo region until 1867. What they did have was German dance hall girls who were known as "hurdy-gurdies" or "gurdies" for short. The Barkerville *Cariboo Sentinel* had a staff poet, James Anderson, who wrote a poem about the gurdies in a Scottish dialect:

Bonnie are the hurdies O!
The German hurdy-gurdies O!
The daftest hour that e're I spent,
Was dancing with the hurdies O!

By today's standards that was a pretty daft hour

indeed. The going rate for one dance was $10. How many dances are there in an hour?

While all this excitement was happening in the remote Cariboo Region, Victoria was growing into a small cosmopolitan city. It was gold, as with San Francisco, that fueled the growth. But unlike San Francisco, it did not explode upon a map; rather, Victoria expanded like a big balloon. When the incipient gold rushers arrived in British Columbia, many decided to stay in Victoria rather than endure the rigorous Cariboo conditions. A visiting Englishman was astonished to find: "...persons who had crossed the Andes, fought in the Crimea, explored the Northwest Passage, seen Peking, ransacked Mexican antiquities, lived on the coast of Africa, revelled in the luxuries of India, wintered in Petersburg and engaged in buffalo hunts on the great plains of North America." The point the Englishman missed is that the Cariboo Region must have truly been rough to dissuade these hardy individuals from going there.

In recognition for his part in the orderly development of British Columbia, James Douglas was knighted in 1863. A year later he retired and the Cariboo Road was completed. It cost a mere $1,250,000 for 383 miles of road. The days of mule trains were over. They were replaced by horses and oxen. Miners now traveled on fast coaches. The unreachable was now accessible. British Columbia came into its own. In 1871 British Columbia became part of the dominion of Canada with the understanding that the transcontinental railroad would be built.

The gold mining boom phase was over by the 1870s, but gold and mineral mining continued on a more even keel and today remains an important part of British Columbia's economy.

Douglas died in Victoria in 1877 and became known as the father of British Columbia. William Dietz lost all of his cash buying drinks and women over the course of many months. He ended his days the old men's home in Victoria, as a ward of the state. Cariboo Cameron set up a business in the East and lost his fortune. Like a wounded animal he returned to William's Creek and died in poverty at Barkerville. A few of the gold rushers became very wealthy, some did well and most spent what they made in the mining country. The real winners were the region, the suppliers and the people who came to settle the land. But what an exciting time it was.

▲
Camels were thought to be the ideal answer to the problem of maintaining pack animals on the Caribou Road during the dry season. Altogether, 21 camels were imported to British Columbia, and for a short time it looked as if the scheme would work. But there were problems with the animals that no one had thought of. This 1888 photograph is of one of the last surviving camels to briefly travel the Caribou Road.

Ships that ventured into San Francisco Bay during the gold rush were in danger of being lost, not from storms or piracy, but by abandonment from passengers and crew who rushed off to make their fortunes. By 1852 there were about 500 ships rotting at anchor.

Battered by a hurricane, the California-bound Comet *off Bermuda in 1852. Rounding the Horn, storms were as severe and could hold up a ship for weeks on end.* ▶

LITH BY N. CURRIER.
SHIP "COMET" OF NEW YORK.
HURRICANE OFF BERMUDA, ON HER VOYAGE FROM NEW YORK TO SAN FRANCISCO, OCTR 1852.
NEW YORK, PUBLISHED BY N CURRIER, 152 NASSAU STREET.
E. C. Gardner, COMMANDER.

▶
Gold miners sod house in Pinas Altas, NM. Building houses like this took advantage of the natural insulation of the earth piled up one side. In the background off to the left is a sluice system.

▶
Front of a gold miner's sod house. Notice the two rockers by the door. A placer miner with two rockers was either very successful or very hopeful.

▶ *A sluice made from native wood. In many cases gold rushers could not afford to by cut lumber and they would fashion their sluices by hollowing out a large tree.*

▶
Arrastas were used to crush gold bearing ore with the help of a large rock and pack animals. The animals were hitched to the long arm on top and were walked in endless circles around the stone wall. Most of the early arrastas used one rock and some like this later 19th century model used three rocks.

▶
When the land gave up all its gold, miners moved on, leaving all their construction behind.

▲ *Earthly gold was born far away in time and space. The Great Nebula in Orion is the remnant of a super nova where gold was created in the fiery death of a giant star.*

CHAPTER 7

COLORADO

The mean elevation of Colorado is 6,800 feet, pretty much what you would expect from a state that has the Rocky Mountains running north and south through its center. There are 55 named peaks that are over 14,000 feet and 1,000 other summits that are over 10,000 feet. Colorado is also known as the mother of rivers. It contains the continental divide of the United States, a watershed boundary in which snow falling on the opposite sides lands in different watersheds. In Colorado there are three such watersheds: one which feeds the Mississippi, one which ultimately reaches the Gulf of Mexico, via the Rio Grande and one which flows westward through the Colorado River to the Pacific Ocean. Adjacent to the eastern slope of the Rockies are the plains, a treeless, grass covered region once called the "The Great American Desert." Occasionally the flat land is broken by isolated buttes and white sandy beds of seasonal streams. It is on these plains that huge herds of buffalo once roamed. So large were these herds that when they stampeded, the ground would shake and all one could see for miles was a brown cloud rushing across the landscape. West of the Rockies is the plateau region, which consists of a series of mesas or plateaus that decline gradually, in a step-like fashion, toward the western edge of the boarder. Cutting deep groves into the mesas are deep river valleys.

THE HIKE

The earliest Indian tribe in this region, about 500 B.C., was called the Anasazi or Basketmakers, so named because of the extraordinary examples of their basket weaving that have survived to this day. In some cases, the baskets have been uncovered almost intact. Several of these baskets were watertight and used for cooking. Baskets were not put directly on a fire, but rather super-hot rocks were tossed into a watery meal until it boiled. The Anasazi are best known for their early examples of stone houses built in the sides of mountain. They were the predecessors of the master cliff dwellers, the Pueblo Indians.

Francisco de Coronado passed through the southeastern corner of the state in his unending quest for the Seven Cities of Gold. Throughout the 16th and 17th centuries few white men came to Colorado, and then only to prospect or to graze their livestock. In 1706, while Juan di Ulibarri was chasing runaway Indians, he made an amazing discovery. To set the record straight and boost his own status back home at Royal Court, he took formal possession of an enormous parcel of unclaimed land for the king of Spain. During the next 100 or so years only the Spanish explored the area and then in the most cursory manner. In 1803 the first American, James Purcell, wandered into Colorado looking to expand his fur trading business. He was followed by Lt. Zebulon

Pike, who in 1806 was busy following the Arkansas River to its source. On the way, Pike discovered the peak which bears his name. John C. Frémont, Jim Bridger and Kit Carson were just a few of the men who explored the region in the following years. Mountain men trapped furs in Colorado until the 1840s when beaver went out of fashion in the East. In the late 1840s small groups of Argonauts passed through Colorado dreaming of California gold. A few years later, in 1850 to be precise, a party of over 100 Cherokees found a little gold near the site of the future city of Denver. They took this as a sign that more wealth was waiting for them in California and hurried on. In 1858 there were not many whites living in the Colorado Territory. There a few Mormons, a few trappers and a few military detachments manning fortified trading posts. Most of the other 3,000 to 4,000 inhabitants were Cheyennes, Arapahoes and other Indians. The real development of the region would not occur until Colorado had its own gold rush.

Rumors of Spanish and Cherokee gold finds in the Colorado Rockies had spread among the California gold rush veterans for years. William Green Russel was not only a veteran of the California rush, but Georgia's as well. He too had heard these reports and in 1858 organized a party to prospect the Pike's Peak region. The group left Georgia and journeyed along the Arkansas River into north and central Colorado. They focused their attentions in a vague area near Pike's Peak. The party panned and rocked and found crumbs instead of the giant golden cookie Russel had led them to expect. The expedition began to crumble in the wake of such disappointment. Russel managed to convince enough of the faithful, or wishful, to set up winter quarters by the mouth of Cherry Creek, while he hiked back to the Missouri Frontier for supplies. Slowly Russel's original group was growing due to the influx of other miners from Kansas. Some of them were part of a small group from Lawrence, Kansas. The Lawrence folk had heard about the gold from an Indian prospector in Colorado. It was not a full blown gold rush because the people who heard the story, for one reason or another, kept the news to themselves.

At the same time a trader was making his way back to Kansas when he stumbled upon the camp. He watched the men mine and traded some items for a small piece of placer gold. This man, whose name is left unrecorded in history, knew nothing of placer gold and believed that this camp was highly successful in their mining endeavors. When he got to the Missouri frontier he showed the gold and told his tale. Almost everyone was eager to believe him, although not for the usual reasons.

In 1858 life in the Missouri frontier towns was quiet, if not downright depressing. Gone were the days of the forty niners. Prosperity was a memory, financial panic a reality. This trader with his small golden nugget could change all that. His news, no matter how small the fact, could start a new gold rush that would be an incredible boon to the frontier towns. It would be an economic shot in the arm. It would be "happy days" all over again. Nothing was going to stop this gold rush, not even the truth.

"A New Eldorado in Kansas" screamed newspaper headlines east of the Missouri. Chambers of Commerce in each of the Missouri frontier towns began a concerted public relations effort to sow the tall tales. They hired special agents, who could be best described as rumor mongers, to spread the word in nearby cities, on the Mississippi and Missouri rivers steamers and on the trains bound for the East and West. These representatives were armed with phoney newspapers full of fictitious accounts of the Colorado gold rush. In order to keep the truth from leaking out, the Chambers of Commerce and newspapers made a preemptive strike. They accused Russel of down playing the enormity of gold strike so he and his friends could keep the gold for themselves. Russel's protests were mere cricket chirps in the greedy storm that was rising. And so the frontier towns got the lucrative gold rush they wanted.

As was the case 10 years earlier, most of the "Colorado fifty niners" had very little placer mining experience. Unlike the California gold rush there was a small number of experienced miners. They had reached a level of professionalism and no matter how unrewarding the search was in the short run, they kept going. In the long run it would pay off in spades because they knew they were reading all the clues correctly. It is the kind of situation where a person doesn't panic, get depressed or quit for lack of an instant pay off. The professionals knew where they were going, which is just as well because few of the other gold rushers did.

"Pike's Peak or Bust" was the rallying cry painted on the canvas wagon tops, and it was leading the gold seekers to the wrong place. The first gold rushers to Pike's Peak discovered that there was no gold to be found within 100 miles of the peak. In early May, as the frontier towns swelled with outward bound gold rushers, there was an opposite wave of inward bound disillusioned gold seekers passing through the towns on their way home. Their wagons now had the Pike's Peak slogan crossed out and under it was hastily scrawled: "Busted by God."

When the forty niners traveled cross-country in their wagons they abandoned useless objects on the way to the gold rush. In contrast, the early "fifty niners" trashed the plains returning home. Among the many discarded articles were optimistic guide books of the region written by D.C. Oakes. He went from hero to goat in several months. He was so disliked by the disenchanted gold rushers that they erected a fake headstone with a rhyme inscribed upon it:

Here lies the body of D. C. Oakes,
Killed for aiding the Pike's Peak hoax.

William N. Byers, publisher of the *Rocky Mountain News*, printed poems like the following:

The gold is there, 'most anywhere
You can take it out rich, with and iron crow bar,
And where it is thick, with a shovel and pick,
You can pick it out in lumps a big as a brick.

Then ho boys ho, to Cherry Creek we'll go.
There's plenty of gold,
In the West we are told,
In the new Eldorado.

At least Byers got the area right, alrhough he received his fair share of criticism and abuse. Some even suggested that he and Oakes be lynched. Someone else pointed out that too many people believed Oakes was already dead and the lynching notion was not taken seriously.

The happy campers who wintered at Cherry Creek gave up looking for gold in the snow buried stream. Instead, they spent their time making plans. Urban planning was first on the agenda. Out of the long, cold nights filled with politics and fueled by rivalries two cities were born, Auraria and Denver. Each city was situated on opposite sides of Cherry Creek, representing two opposing viewpoints on certain issues that have since been forgotten. Denver was named for James W. Denver, governor of the Kansas Territory, which comprised of the future states of Kansas and Colorado. In 1860 Auraria and Denver were consolidated into one city.

Pike's Peak wasn't a total bust. Julia Archibald Holmes became the first non-Indian woman to climb the mountain in 1857. She accomplished this feat at a time when women, good women; women of culture and breeding; women of fine moral character; women who knew their correct place in society, certainly didn't do anything so bold. It was a time when the idea of a woman climbing a mountain was as far fetched as a man walking on the moon. But what can you expect from a century when whole sections of a country still had slavery as an institution.

Julia's husband, James Henry Holmes, of Lawrence, Kansas was a free-stater. In 1856 Holmes escaped with his life when a group of proslavery Missourians raided the free-state stronghold where he worked. These were raids where property damage was the least expected outcome by the pro-slavery thugs; they wanted to see bodies, lots of dead, free-stater bodies. It was an era referred to as "Bleeding Kansas," where people died in the struggle to determine the final slavery status of Kansas when it was finally admitted to the Union.

Holmes hungry for revenge, took part in several free-state raids in the homeland of mortal enemies, Missouri. Not one to miss the chance for making a few dollars, Holmes stole a few horses belonging to the pro-slavery Missourians. A month later he was arrested for horse stealing by order of John Geary, the Kansas territorial governor. While being escorted by the U.S. Army to his trial in Missouri, Holmes escaped and fled to Emporia, a little frontier town on the territorial boarder.

Meanwhile, early in 1857, Julia Holmes was in Lawrence attending a meeting of the free-state legislature. At this particular assembly there was little or no political discussion. All the attendants could think about were rumors of a Delaware Indian named Fall Leaf who lived on a reservation outside of Lawrence. He said he went west, into the Rocky Mountains and found gold. Fall Leaf even brought some samples back with him. Although no one had seen any of the specimens, they all believed the stories. The legislature broke up sooner than expected and many of the participants feverishly made plans to go to the Colorado Territory. Julia rushed to meet her husband and tell him the news. They agreed to join one of the Lawrence parties camped a short distance from Emporia. It was a small group of 47 gold rushers led by a butcher, John Easter. No one in the party knew much about the Rockies or how to get there. They just blundered their way west.

On July 6 the band camped near the future site of Colorado Springs, at the foot of Pike's Peak. Most of the men, James included, went searching for gold. But four men decided to climb Pike's Peak, for the classic reason, because it was there. Five days later they returned to the camp. Their enthusiastic stories of the climb and the wonderful sights they had seen inspired Julia to duplicate the feat with her husband.

On August 1 Julia and James Holmes set out to conquer the mountain. They carried over 50 pounds of supplies, clothing and utensils between them. Julia shouldered 17 pounds in her pack and James hauled about 35 in his. Julia packed one extra item, a book of poems by Ralph Waldo Emerson. Julia did not wear a dress for the climb, instead she was very practical in her thick moccasins, black pants and tan shirt. Most women of her day would have been shocked at her apparel, no matter what the circumstances.

For three days they trekked through canyons and climbed steep slopes that were occasionally treacherous from loose gravel. On the third day, within sight of the summit, Julia and James camped in a cave they called "Sun Dale" for the evening. The next day's climb would take them to the top of Pike's Peak. As a bright red dawn broke on the mountain Julia and James made their final ascent. They past the timberline; the ground was sprinkled with large patches of snow and tiny blue flowers. Shortly after noon they made it to the summit and Julia Archibald Holmes became the first white woman to climb Pike's Peak.

The couple spent an hour enjoying the incredible vista. Julia read Emerson's poem titled "Friendship" and wrote a letter to a friend describing her feelings: "Just being up here fills the mind with infinitude, and sends the soul to God." They returned to the gold rush camp three days later.

For Julia, James and the rest of the Lawrence party, there was no gold to be found in Colorado. The party broke up and went their separate ways. A small group, the Holmes included, wintered in Taos, New Mexico. Julia and James remained in Taos rather than going back North and searching for gold. James became territorial secretary and Julia became the New Mexico correspondent for the *New York Tribune*. In most tales of love and adventure we expect the "And they lived happily ever after" ending, but this was real life and Julia and James divorced after a few years. Julia went East and became secretary of the National American Woman Suffrage Association, located in Washington, D.C. There was another first added to her list of accomplishments. She became the first woman member of the U.S. Bureau of Education.

However, there was gold in Colorado, but it wasn't easy to find unless one had experience. Twenty-eight-year-old George Jackson, former California gold rusher, had the necessary requirement. He had already done more in his short life than most people could possibly do in several lifetimes. His formal education only went as far as completing elementary school. Jackson's informal training made him one the best trappers, riders, marksman and hunters in the West. Throughout his life, no matter where he went or who he met, he was always well-liked and considered the consummate frontiersman.

He left home at an early age and lived with his cousin, the famous Kit Carson, in Santa Fe. He was one of the first to arrive in California for the gold rush. Sutter offered to sell him 100 acres of land near Sacramento, but Jackson declined the proposal and returned to Missouri in 1852. Five years later he was on the move again, acting as a guide for a small company of U.S. Army troops sent to put down the Mormon revolt in Utah. He spent the next year and half hunting and trapping near Taos. In July 1857 he appeared with two friends and a bundle of goods at the Cherry Creek Indian trading post.

In January 1859 Jackson had discovered a placer gold deposit on a fork of Clear Creek, about 30 miles due west from Denver. Because he was low on supplies he could only pan a few ounces before he and two friends had to leave. In the spring he returned with a party from Chicago. The group consisted of 22 men who were not only workers but also investors. It was called the Chicago Mining Company whose sole purpose was to develop Jackson Bar and get on with the business of gold mining. After arriving at Clear Creek they immediately converted their wagons into sluices. The first day they made $1,900. A few days later another $5,000 was found. Word of the Jackson Bar gold strike spread throughout Colorado. Miners flooded the Cherry Creek region and the earth yielded up its treasure. In 1920 it was estimated that over $100,000,000 in gold was unearthed from the area George Jackson discovered.

John Hamilton Gregory was the opposite of Jackson. Where Jackson was young, Gregory was middle aged, 40 to be exact. Gregory grew up in Dahlonega, Georgia during the gold rush. It was here that Gregory learned the business of gold mining. Although he knew how to read and write, he never went to school. As an adult, his principal occupations were drinking and gambling. When Gregory spoke, each sentence he uttered was sprinkled with profanity. In his later years he wore a battered felt hat and red shirt which were hardly ever removed. That this man participated in the 1858 Fraser River gold rush is hard to believe, yet he did. In January 1859 he wandered into the Colorado Territory and panned the streams near Pike's Peak. It is here that he met George Jackson and made a very fortunate mistake.

Low on food, Jackson and his two friends were hiking out of the mountains to the town of Arapahoe when they met the disheveled Gregory. They camped together for the night and Jackson confided in the older man the location of the yet to be developed Jackson Bar. The next morning they went their separate ways. Nursing a hangover, all Gregory seemed to remember is that he should go up to Cherry Creek and stake a claim. Somehow Gregory took the wrong branch of the creek and ended up traveling further north to what is now the town of Blackhawk. Before the weather changed and it began to snow heavily, Gregory managed to pan some gravel which revealed a few gold colors. He too made his way toward Arapahoe to spend the rest of the winter. In the early spring Gregory wandered around the mountains seriously thinking of all the gold dust he did not have and returning home without ever having made that one big gold find. Then he met the Wall party and his life took another surprising turn.

The Wall party hailed from Indiana and was comprised of David K. Wall and nine of his friends. The young men, barely in their twenties, had come West to hunt game and didn't even know there was a gold rush going on in Colorado. Their adventures took them across the Plains and toward the Rockies. Near the mountain chain they met a group of disenchanted gold rushers returning home. In a fearful tone the leader told David Wall, "Go back, for God's sake. They are starving in Denver and they will kill you for your provisions." Being carefree and adventuresome men, the Wall party cleaned their guns and headed straight for Denver. When they arrived at Denver they found the only thing the inhabitants lacked were delicacies. So Wall and his companions traded dried fruits for the supplies they might need in the mountain country.

They stayed a few days and headed north into the snow covered Rockies. The young men came upon, "a ragged and sorry looking specimen wearing half an overcoat." The wretch was none other than John Hamilton Gregory, who told them a hard luck story laced with morsels of actual fact. After the Wall party fed him and gave him clothes, Gregory spoke of his find on the north fork of Cherry Creek. Then he made a deal with the Wall party. Gregory said "Boys, if you all want to put up grub and transportation against my experience as a miner, we'll go up the creek and we'll get gold." The group cheered and cemented the transaction with a drink.

Maybe Gregory had another hangover that morning for he led them not to Cherry Creek but Chicago Creek, and from there to Clear Creek. They panned from stream to stream without finding gold. To the Wall party this was all a fabulous enterprise, something they never dreamed of doing. They didn't know enough to get discouraged. On May 6, 1859 Gregory went a little further up the creek to an area where the water met the side of a hill and panned $5 worth of gold. This was the place he told the group, and they made camp after naming the site Gregory Gulch. Subsequently it turned out that this was the site of a particularly rich quartz vein. In the months to come the area was named Gregory Diggings and flooded with 5,000 miners, all finding gold. During the next two years Gregory Diggings, as the whole area was called, yielded several million dollars in gold.

June marked the coming of the three newspapermen to Gregory Gulch. They were Henry Villard of the *Cincinnati Daily Commercial*, Albert D. Richardson of the *Boston Journal* and Horace Greeley of the *New York Tribune*. Everyone knew who Greeley was, but Richardson would not attain national fame until the publication of his book, *Beyond the Mississippi*, which detailed the Western trip the three of them took. Later, Villard became an extraordinary correspondent and newspaper publisher as well as a great Western railroad magnate, but to the miners at Gregory Gulch the only one who really mattered was Greeley.

From the Missouri frontier, the party made its way to Colorado by stagecoach. Throughout the ride the reporters found themselves drenched in unremitting sunlight during the day, and alkali dust both day and night. Greeley had heard rumors to the effect that Denver was practically a ghost town, and many gold rushers were nearly starving for want of money to pay for food. These were similar to the stories that the Wall party heard from failed miners who left Pike's Peak. Greeley and company wanted to see the situation for themselves. Even though Cherry Creek was almost mined out, Denver was growing by leaps and bounds when Greeley and his friends arrived. News of Gregory Gulch reached Greeley here, and it was off to the wild mountain country for the newspapermen. Up to this point the three reporters had not seen any of the gold produced by the rush. The men at Gregory Gulch were going to rectify this.

The night before the coming of the three newspapermen the miners spent their time thinking of the best way to please their visitors. They came to the conclusion that it was very important to impress Greeley about the richness of the gold in Gregory Gulch. Soon a crafty plan emerged that combined trickery and the foibles of human nature; not even one so smart and cultured as Greeley would guess he was had. The plan included salting a recently dug hole with gold in an unclaimed area.

The distinguished guests arrived with the morning sun and taken on a tour by the leaders of the camp. Many of the amenities Greeley's readers were used to were missing at Gregory Gulch. He noted the primitive shelters made from earth and pine branches alongside hastily erected ragged tents. Meals were cooked and served in the open and the miners ate on the ground without benefit of chairs, tables or cloths. The scene looked and smelled very much like the California gold rush a decade earlier. After seeing all the claims, hearing about the brief history of the gulch and examining recently mined specimens of gold, Greeley got the itch, just as the miners expected him to. "Well now, Gentlemen, I would like to see a hold and dig out and wash some of the dirt myself. I want to actually see the gold taken from the dirt," said Greeley.

The wise three newspapermen were taken a little ways up the gulch to the salted hole. They were told that work had not progressed this far and, although the site was started, no one had really worked the hole yet. Greeley was given a shovel and he dug up some dirt. Next he was given a pan and showed how to separate the heavier golden elements from the dirt. When he panned to the bottom of the debris he found gold dust. Excited, he dug and panned some more. Again he found gold. Greeley repeated the process three times until he grew tired. Gathering his gold dust into a bag he enthusiastically proclaimed, "Gentlemen, I have examined your property with my own eyes and worked some of it with my own hands and I have no hesitation in saying that your discovery is what it is represented to be, the richest and greatest in America."

In a June 11 dispatch widely distributed to newspapers around the country, the three reporters, in a joint article, told how miners were practically picking gold everywhere they went. To be fair they went to great lengths to warn of the hazards of gold mining and the chance of failure. The dispatch ended with the caution, "...We beg the press generally to unite with us in warning the whole people against another rush to this gold mine, ...a rush sure to be followed like a stampede, but one far more destructive of property and life." This is like telling a child that the cookie jar on top of the refrigerator is full of great cookies, but don't go up there because you might fall and hurt yourself. No one is surprised when the child climbs on top of the refrigerator to get the cookies. No one was surprised when more gold rushers came to Colorado to seek their fortunes. The only miners who did not arrive in large numbers were the Californians who were rushing to British Columbia instead.

Other gold sites were discovered by the fifty niners. They had names such as Tarryall and California Gulch. Tarryall got its name from a conversation between two prospectors who were part of a large party searching for gold in the South Park area of Colorado. They had reached the banks of a small creek and one of the members said, "Let's tarry here." He was answered by another who said, "Yes, lets tarry all." The phrase became a name and Tarryall stuck, except for a short while when the site was overrun by gold rushers who found that all those who came before had staked all the available claims. Then it was called Grab-all.

In April 1860, west of South Park, a party of Iowans led by L.P. Jones met a party of Georgians led by Abe Lee. They were both prospecting for a rich site in 4 feet of snow. Lee and Jones decided to pool their respective party's resources so both groups could cover more ground and arranged a system of signals. If anyone in either party found a rich placer they would fire their rifles and set a bonfire to mark the spot. Abe Lee found a likely looking area and panned a few colors. Thinking this might be a good site he dug a little deeper and produced another sample which he panned. Lee could not believe his eyes as he stared at the bottom of his pan. It glittered like the rising sun. "Boys," he said to his nearby fellow Georgians, "by the Almighty, I've got California in the bottom of this here pan!" Dozens of gun shots rang through the mountains and bonfires lit the skies. California Gulch was named and it became the largest placer district in Colorado.

According to contemporary figures, about 100,000 left the Missouri frontier at the beginning of the Pike's Peak rush, but only 25,000 actually reached Colorado. By the end of 1859 only 4,000 of the initial prospectors were engaged in mining. Throughout the first year other miners, who were not counted, drifted in and out of Colorado. Between 1858 and 1870 over 100,000 people lived in Colorado, but the number of gold rushers actually there at any given moment was closer to 30,000. They left in their wake dozens of gold mining towns such as, Golden, Blackhawk, Nevadaville, Empire, Gold Hill, Mill City and the most famous in its day, Central City.

Since most of the real gold was in the form of quartz lodes, mining became very technical within a short period of time. This does not mean that there were no other placer deposits to be found. The placer deposits already found were really indications of nearby quartz lode deposits, and to mine the gold more capital and labor intensive means needed to be employed. Many miners became hourly wage earners. According to Villard, half the miners he saw in 1959 were hired hands. Ore crushing and grinding machines were used early in the rush to extract the gold from quartz lode deposits, and ore operations usually processed a lot of material to get a small amount of gold. One fact which would have a greater impact in later years emerged from the gold rush Colorado possessed a mineral belt that stretched diagonally across the state, from Boulder County in the northeast to San Juan County in the southwest.

The first Colorado gold rush lasted from 1858 to 1867. The peak years were 1862 and 1863, in spite of the fact that the Civil War was draining men away from the business of mining. Each of those years yielded about $3,500,000 in gold. All told the total gold mined came to about $25,000,000, not a large tally compared to the early years of California. Worse yet, the ores that were mined at 100 feet or more were less rich and found in a chemical combination of sulfides. Techniques using mercury just wouldn't work with gold found in such a state. In fits of frustration the miners called them refractory or rebellious ores. It was this lack of spectacular gains coupled with hard to refine refractory ores that deterred gold rushers from staying longer and settling in this mountainous territory. What Colorado could not offer in terms of gold it offered in terms of land. Settlers slowly filtered into the territory as the miners faded away.

Things turned bad in 1864. The previous year had seen many of the miners unable to finance the more improved and scientific processes needed to extract gold from the refractory ores. More and more capital was raised from the capitalists in Boston, New York and Philadelphia. In many cases these fund-raising efforts resulted in an outright sale of the claim. These new owners, who were thousands of miles away, believed more powerful machinery to refine the gold was needed. But the orders for these machines were delayed due to the fact that most of the North's industrial capacity was being used to supply military materials for the Civil War. The recently bought mines were unworked and resulted in regional unemployment. As if this wasn't bad enough, the winter of 1864-65 was unbelievably cold and the next spring brought an invasion of grasshoppers that devastated the new farms along the rivers. But by far the worst event in those dark days was the Indian wars.

In the early days of the gold rush the Southern Cheyennes and Arapahoes settled very peacefully in Denver. It was a town built because of gold, and in its first few years gold remained the primary business. In 1859 there was a sizable Arapahoe village in the middle of the new city. There was much trust between the whites in the city and the Indians' encampment. In fact, in the summer of 1860, the Arapahoe warriors left to fight the Utes and felt safe leaving there women and children unprotected in Denver. An 1861 treaty began to change all that.

From the beginning, the treaty was unpopular with the Indians, because in its provisions the Southern Cheyennes and Arapahoes gave up much of their territory. Worse still, a clause was construed by government officials allowing the railroads, when built, to pass though what remained of Indian land. The railroads meant the establishment of permanent white settlements that would compete for the available land, land that was needed for grazing, hunting and growing native crops. The Indians had already seen examples of this threat in the Eastern plains when their principal game, the buffalo, was slaughtered in such large numbers that they almost became extinct.

Trouble brewed for three years as the Indian population of the two tribes forced their own chiefs to repudiate the treaty. Politicians in Colorado viewed these rejections as hostile acts. The Indians were infuriated by the influx of white buffalo hunters to their lands, needlessly killing their primary source of red meat just so the white man could wear fashionable skins. But it took the coming of the Rev. J.M. Chivington, colonel of the Colorado volunteers, to end the peace.

In the spring of 1864 this self-righteous reverend reported that a group of Cheyennes had rustled some cattle from a government contractor's herd. No one really believed the story since new settlers were constantly losing cattle and covering up their own ineptitude by blaming Indians for the loss. But Chivington did not wait for government approval or action, he immediately took his volunteers and attacked Cheyenne villages. What was especially galling to the Cheyenne was the fact that Chivington and his troops chose to raid the villages when most of the men were away hunting. One bad turn deserves another and the Cheyenne attacked the homesteaders, which was particularly surprising to the settlers because they had no idea there was an Indian war going on.

John Evans, Colorado's second territorial governor, was an exceptional man. Raised as a Quaker, in his adult years he switched religions and became a Methodist. He was also a doctor, medical teacher, founder of hospitals, the founder of Garrett Biblical Institute and Northwestern University. At the same time, Evans made a small fortune in Chicago real estate and railroad promotion. In his spare time he was active in Chicago civic affairs and the nascent Republican party. Lincoln selected Evans to be territorial governor for two reasons: Evans was a prominent leader from his own state and the Methodists were complaining that he had neglected the denomination in his other political appointments. It was a good choice and Evans would make a deep mark in the development of the state. In 1864, however, he had Indian troubles.

As many a good leader before and after him, Evans used his energy and personality to bring the warring parties to the conference table. The Cheyenne were naturally suspicious of the governor's intentions, so they sent their two best leaders, Black Kettle and the esteemed 70 some odd year old war chief White Antelope, to negotiate in Denver. An interim agreement was reached allowing the Cheyenne camp about 30 miles from Fort Lyon while the negotiations continued.

Chivington had another agenda. He was overjoyed to hear the Cheyenne had situated their village in such an accessible area. Dreaming of military glory, he called himself by his volunteer rank of colonel and planned a gory massacre to secure his fame. Having the guarantee of military protection from the U.S. Army at Fort Lyon, the Cheyenne suspected nothing. Chivington gathered about 800 men and under the cover of darkness they rode to the Indian camp. His last order to his men before the onslaught was, "Kill and scalp all big and little; nits make lice." With a whoop and holler Chivington's "boys," as he frequently referred to them, attacked the sleeping Indians.

The Indians didn't have a chance. Over 350 Cheyenne and Arapaho Indians were killed by Chivington's men, close to 200 were women and children. Black Kettle tried to stop the slaughter by displaying both the American flag and white flag. Only the Indians who were fleeing for their lives noticed his act. A woman tried to escape and was shot down. As she lay wounded on the ground, passing volunteers shot her again. This woman was Black Kettle's wife. She was shot nine times and lived. Black Kettle escaped, but White Antelope just stood in front of his lodge singing, "Nothing lives long, except the earth and the mountains." It was his death song. He was ripped apart by a fusillade of bullets.

Chivington and his "boys" put their war trophies, the scalps and severed limbs, on display at a theater in Denver. The first crowds who came to the gruesome exhibit cheered, but later throngs were upset and some were appalled by the macabre show. Kit Carson was infuriated. He called Chivington and his troops cowards and dogs. Many of the settlers had an uneasy feeling that this victory did not portend well for the future.

The Plains went up in flames that summer. The war was carried to every settlement. In July and August 117 men were killed and an unrecorded number of the dead settler's women and children were carried off as captives. Work on the Union Pacific Railroad was halted because of Indian raids. Colorado found itself nearly isolated after all the telegraph lines into the state were ripped down and stagecoach way-stations burned. For four years the war ensued. It finally ended in 1869, when Gen. George Armstrong Custer won the last battle. A government commission studied the conflict and concluded that Chivington's raid, "...scarcely has its parallel in the records of Indian barbarity....No one will be astonished that a war ensued which cost the government $30,000,000 and carried conflagration and death to the border settlements."

Colorado went on to prosper. The 1870s saw a period of rapid development: the Indians were removed to reservations, new and more scientific methods of smelting refractory ores were devised, the first railroads came, farming increased and new agricultural towns were founded. In 1876 Colorado was admitted to the Union and a year later fortune shined on the new state with the discovery of silver at Leadville. A new rush was on and this time the metal shone like the moon. Besides silver, the broad diagonal mineral belt also contained rich deposits of lead, zinc and copper.

Of the winners and losers it was the same old generalization. The early miners got the glory and the large industrialized mines got the cash. Famous participants included: Henry Morton Stanley, a world famous journalist and explorer, who found Doctor Livingston in Africa; George Mortimer Pullman who prospected, opened a hardware store and perfected his idea of railroad sleeping cars by adapting the design of the two tiered miner's bunks and Nathaniel P. Hill who built the first commercially successful smelter in America.

John Hamilton Gregory, the man who started the real Colorado gold rush, sold his share in Gregory Gulch for $21,000 and went back to Georgia. He had dreams of making his wife a fine lady and bringing up his children surrounded by great wealth. Before long he spent all his money without realizing this vision. Gregory returned to Colorado to prospect for gold and was never heard from again.

The dashing George Jackson left Colorado to fight for the Confederacy. Jackson was promoted to lieutenant colonel after he captured the United States cutter, *Harriet Lane*. After war he returned to Colorado and continued prospecting. His search for gold took him to both New Mexico and old Mexico. Although he discovered minor gold sites, he never made another spectacular find. In 1890 he sold the Saratoga Mine in Colorado for $40,000 and retired to Chipita where he accidentally shot himself in 1897.

In the 1890s there was one more Colorado gold rush. Silver had been the primary ore mined for the last 20 years, but now Leadville, the bonanza silver camp, was inactive and the gold mines near Central City were in a slump. Added to the silver troubles was the Panic of 1893, a very complex financial and political crisis. The immediate cause was the inability of the United States government to maintain a $100,000,000 gold reserve needed to assure redemptions of Treasury obligations in gold. The immediate effect of the panic was the repeal of the Sherman Silver Purchase Act. This law required the secretary of the Treasury to buy $4,500,000 of silver each month at the full market price. Alarmed at the falling prices of silver, the mine owners put pressure on Congress to adopt the law. It was passed in 1890. When the panic struck, it was widely believed that the silver purchase policy was responsible for the the unprecedented drain on the Treasury. As a result of the repeal of the Sherman Silver Purchase Act, the price of silver dropped faster than gold to the bottom of a prospector's pan, and as a result of the sharp fall in prices most of Colorado's silver mines were wiped out. Agriculture, tourism and cattle ranching were the only businesses holding Colorado's economy together. Then, gold was discovered at Cripple Creek on the western slope of Pike's Peak, and instantly the economic picture changed.

Cripple Creek was a grass covered depression surrounded by a cluster of small pine covered hills. Cattle grazed in this dimple and prospectors walked through the landscape for years without anyone suspecting there was a rich lode of gold under their feet. It was an unusual setting to find gold ore for more than the obvious reasons. Most gold strikes were found in rugged mountain country, whose topography is crammed with fast running creeks, dry stream beds and gulches. Cripple Creek was located in the throat of an extinct volcano. Its gold ore was in the unfamiliar form of gold tellurides. This is one of the rare instances in which gold is found in a chemical combination with another element, tellurium. But there was enough free gold flakes and dust to start a rush. It turned out to be one of the richest finds in the world.

By 1895 the mining companies at Cripple Creek made great strides in finding solutions to the problem of extracting telluride ore. Modern chlorination plants and cyanide mills sprung up overnight. So rich were the mines that most of them were locally owned and able to finance their own development. Even though most of the miners were hired hands, it became common practice to lease parts of the mines to the workers. This custom may have been initiated partially as result of the Western Federation of Miners' strike of 1894. At the core of dispute was the mine operator's lengthening of the miner's day from eight to nine hours. Troops entered Cripple Creek not to break the strike but to enforce order and bring about a negotiated peace. This highly unusual late 19th century event was due to the fact that Davis Waite, a Populist, was governor. In 1904 the miners at Cripple Creek staged another strike in support of the smelter's attempt to gain an eight hour day. Now James Peabody was governor and he was no Populist. Striking miners were intimidated by the governor's troops and the strike leaders were sent packing.

Colorado surpassed California's gold production in 1897, and in 1900 the state accounted for over a third of America's total output of gold and silver. And so, in spite of the initial shortcomings, Pike's Peak was worth the hike.

▲
Placer mining in the Gregory District in 1859.

CHAPTER 8

NEVADA: THE COMSTOCK LODE

Known as the "Sagebrush" or "Silver State," most of the Nevada, with the exception of the north and southeast corners, lies within the Great Basin, a vast tableland between 4,000 and 5,000 feet above sea level. It is not a plain such as the one found in eastern Colorado, but rather a plateau whose flatlands are broken by many buttes, mesas and isolated north and south mountain ranges. These mountains are anywhere from 1,000 to 7,000 feet above the level of the plateau. The Nevada the skies are clear and blue almost every day of the year; the mean annual rainfall varies from 3 to 12 inches. Despite the dry climate there is no part of the state, not even the desert, that doesn't have some form of vegetation.

EVERY LODE HAS A SILVER LINING

Nevada wasn't discovered the way California and British Columbia were. It was traversed. The first recorded person of European descent to pass through Nevada's modern borders was Francisco Gracés, a Franciscan monk who was on his way to California. The year of his brief visit was 1775. In 1825 Peter Ogdin and a group of Hudson Bay Company trappers traveled through Nevada entering by from the north and leaving somewhere in the southwest corner. On their way they discovered the Humboldt River. When Jedediah Smith made the first recorded journey from the Mississippi to the Pacific by the central continental route in 1827, he passed through Nevada on his return trip. John C. Frémont and Kit Carson were the first persons who did any serious exploration of the state between 1843 and 1845. It is interesting to note that Carson's explorations in Nevada, rather than his other adventures, prompted the native population to name the Carson River, Carson Sink and Carson City after him.

Nevada became a United States territory, along with California, with the signing of the treaty of Guadalupe Hidalgo at the end of the Mexican War in 1848. At first it was part of California and named the Washoe country, after the Washoe Indians. It was in the Washoe country that gold was discovered. In 1850 most of the state was incorporated into the Utah territory. Nevada was admitted to the Union in 1864.

Placer gold was the first type of gold discovered in the Washoe country, but at the time not many prospectors wanted to rush to the inhospitable land. Members of the ubiquitous disbanded Mormon Battalion, on their way to Utah, were probably the first to find gold dust in the region in 1848. It took a year for the news to spread among the other Mormons in Utah. In 1850 a group of Mormons bound for California stopped and panned meager amounts of gold dust from Gold

Canyon. After three weeks of prospecting they were disillusioned with their finds and continued on their journey. For the next several years, up until 1857, California miners trickled into Nevada hoping to find the source of the placer gold and drifted away disappointed. Most of the Gold Canyon prospectors placer mined, a technique that was rapidly becoming obsolete in California. The days of easy pickings were over and more technical methods were being employed. The Grosh brothers knew a bit more about mining than those who came before them and were not as easily deterred.

It was either 1851 or 1853 when Hosea and Allen Grosh arrived in Gold Canyon. Sons of a Pennsylvania clergyman, they were better educated than most miners and had they done their homework by studying geology and mining before they traveled off to parts unknown. It was said that they had a partner, a Mexican silver miner by the name of "Old Frank," but there was no record of what became of the man when the brothers died. For the next few years the brothers found small, encouraging amounts of gold, just enough to keep them going. In 1856 they discovered the silver lodes that honeycombed the mountain and spent several months mapping them. In a letter to their father, dated sometime in November 1856, the Grosh brothers wrote: "We found two veins of silver at the forks of Gold Canyon, ...One of these is a perfect monster." They had learned something no else even suspected, the slopes of Mt. Davidson were richer in silver than in gold. A year later they felt they had prospected and mapped the area enough to raise capital and develop their claims.

Unfortunately, this story does not have a happy ending. Hosea punctured the arch of his foot with a pick. The wound became infected and progressed into a nasty case of blood poisoning. Lacking 20th century antibiotics to combat the infection, he died. This left Allen to carry on alone. He decided to walk all the way to California, with his maps and ore samples weighing heavily on his back, through the snow and treacherous mountains. He succumbed to frostbite on December 19, 1857. In later years Hosea and Allen's father claimed that his sons were the true discoverers of the Comstock Lode. Although he had his sons' letters to prove his assertion, no one ever believed him. Nothing was easy in the early Nevada mining days. Even the Comstock Lode had to be discovered twice!

James Fennimore, alias James "Finny," alias "Old Virginny" was called other names during his drunken, womanizing life, but these were the two that stuck. It is easy to understand how he came to be known as Finny, a convenient shortening of Fennimore. Old Virginny, on the other hand, was a name the Washoe country locals called him because of his love for his home state of Virginia. When Old Virginny was on one of his frequent drunken binges, he would bore any nearby listener to tears with stories of his former home state. He wasn't a bad man, only a generous drunk who had simple tastes. Despite his weakness for alcohol, Old Virginny was the best judge of placer ground throughout the Washoe country. He had been prospecting the area since 1851.

In the fall of 1858 Old Virginny was on a hunting trip when he noticed a flat topped hill near Gold Canyon. It looked as if it might contain some promising gravel. He made a mental note to investigate sometime in the future. The future came in January of 1859. Old Virginny returned to the hill with three of his friends. Because it was winter everyone else had quit their placer mining efforts and holed up in nearby Johnstown until the spring thaw. At first the group found trace amounts of gold dust, not a fortune but enough to name the small rise Gold Hill. Each of the four men took out claims of 50 feet and began to dig. Soon the hibernating miners from Johnstown braved the cold to inspect this new find. They were not impressed. Only four other men bothered to stake claims on Gold Hill. The men in Old Virginny's party were not deterred by their fellow miner's skepticism and continued digging whenever the weather permitted.

In early April they began to dig a hole 10 feet deep and several feet wide. As they dug deeper into the dark, heavy dirt of the hill, the gold recovered from the gravel got richer. But at 10 feet they found a vein of brick red, decomposed quartz loaded with gold. The men, unknowingly, had stumbled upon Old Red Ledge, a vein of the Comstock Lode itself. Yet only a handful of miners took out more claims on Gold Hill. At the time, there was no excitement in Washoe country, and no one thought the find was important enough to mention to the outside world.

Gold Canyon and Gold Hill were on the southern slopes of Mt. Davidson. Peter O'Riley and Patrick McLaughlin decided to try their luck on the northern slopes of the mountain, at a spring high up in Six Mile Canyon called "Old Man Caldwell's." Their goal was simple, find enough gold to raise a $100 grub stake and move on to richer gold country. The site they chose yielded about $4 a day. Although this was not exactly what they expected, they kept prospecting. Soon another problem presented itself. If they were going to continue placer mining the claim they would need more water for their rocker than the spring could provide. They set out to dig a reservoir. As the sun rose on the morning of June 10, 1859 O'Riley and McLaughlin were digging their reservoir when they came upon the same kind of dark, heavy soil they had seen at Gold Hill. They washed out a pan of dirt and it was the same old story. There were gold flakes and dust at the bottom of the pan as bright and golden as the sun. They had uncovered the tip of the famous Ophir vein of the Comstock Lode.

But what about the man that the lode was named after, Henry T.P. Comstock? Henry Comstock was a Canadian who earned the reputation of being a lair, a cheat, a claim jumper and the laziest man in all of Johnstown. He was so lazy that he was nicknamed "Old Pancake," because he would not even make his own bread and when he did it was so bad it would not even rise. It was as flat as a pancake. Old Pancake did have one survival skill common to men of his ilk, a clever mind. He could talk a rattlesnake out of its rattle.

A few days after O'Riley and McLaughlin's amazing discovery they had a visitor at the site. Old Pancake came riding up to Old Man Caldwell's to check out what the two miners had found. He was a sight atop a broken down mule that was so swaybacked Comstock's feet dangled in the sagebrush. Quickly, he ran his fingers through the day's take in dust and flakes, which was worth about $300. Without a moment's hesitation he announced that he and Old Virginny, along with "Manny" Penrod, had bought the mining rights to the spring from Caldwell the previous winter and he wanted his fair share of the gold. They argued about Comstock's claim for a whole day. In the end Old Pancake, true to form, won his point. Comstock, Old Virginny, Manny Penrod and a fifth man, Joseph D. Winters, were added to the list of partners. Poor Old Virginny truly lost in this deal. Old Pancake bought Old Virginny's share out for the enormous fee of $40, one blind horse and a bottle of whiskey. Old Virginny later complained that his horse was worth $60,000, but he could not afford a saddle.

The men settled down to work at the Ophir mine. O'Riley, McLaughlin and Penrod did the mining, while Comstock did the supervising. He would urge them to work harder on "my mine" and be sure not to lose "my gold." For a while everyone ignored "Old Pancake," but human nature is a funny thing. If a statement is repeated long enough and loud enough, people start to believe some of what has been said. Soon his claim of ownership had an effect on the other miners in the area. They started to call the claim Comstock's Lode. Later this became the Comstock Lode.

After many days of successful mining, the site averaged $300 a day. On June 11 the gold rich dark soil ran out and now the men found a heavy bluish quartz in its place. This was a particularly strange lumpy ore that often clogged their rockers. Not knowing what to with the blue debris the miners threw it away until it formed a large mound. It was Penrod who had the idea that the ore they were dumping might in itself be valuable. So the miners took the precaution of filing a quartz claim in addition to their placer claims. At the same time Manny Penrod sent several ore samples of the bluish quartz to Nevada City, California to be assayed. The answer Penrod received was incredible. The heavy blue ore turned out to contain 75 percent pure silver and 25 percent gold. Its value at contemporary market prices was almost $4,000 a ton!

The Washoe country was famous for veins of gold rich ore that played themselves out after six months to a year of intensive mining. The partners sold their shares in the Ophir and other claims believing that in time the veins would give out. For their efforts Penrod got $3,000, McLaughlin $3,500, Comstock $11,000 and because he held out the longest O'Riley received $40,000. The buyers included Judge James Walsh of Nevada City, California, who raced across the Great Basin plateau attempting to beat the express that carried the results of the assay back to Manny Penrod. Another buyer was George Hearst who was on the verge of building an immense fortune with the purchase of a sixth of a share of the Ophir mine for $450.

Records indicate that the original owners had a high old time with their money and died penniless. Comstock committed suicide during what appeared to be a fit of depression. He had wrongly assumed that fate would continue to smile favorably upon him. This was a common misconception held by many gold rushers who made the big find of a lifetime. After his wife ran off with another prospector, and after he lost his money in several bad business deals, he wandered throughout the West searching for another big strike. In 1870, somewhere near Bozeman City, Montana, he ended his search by putting a loaded revolver to his head and pulling the trigger. Old Virginny just kept boozing his life away. Although he was cheated out of his share of the Ophir, he did well by his find on Gold Hill. In July 1861, in the

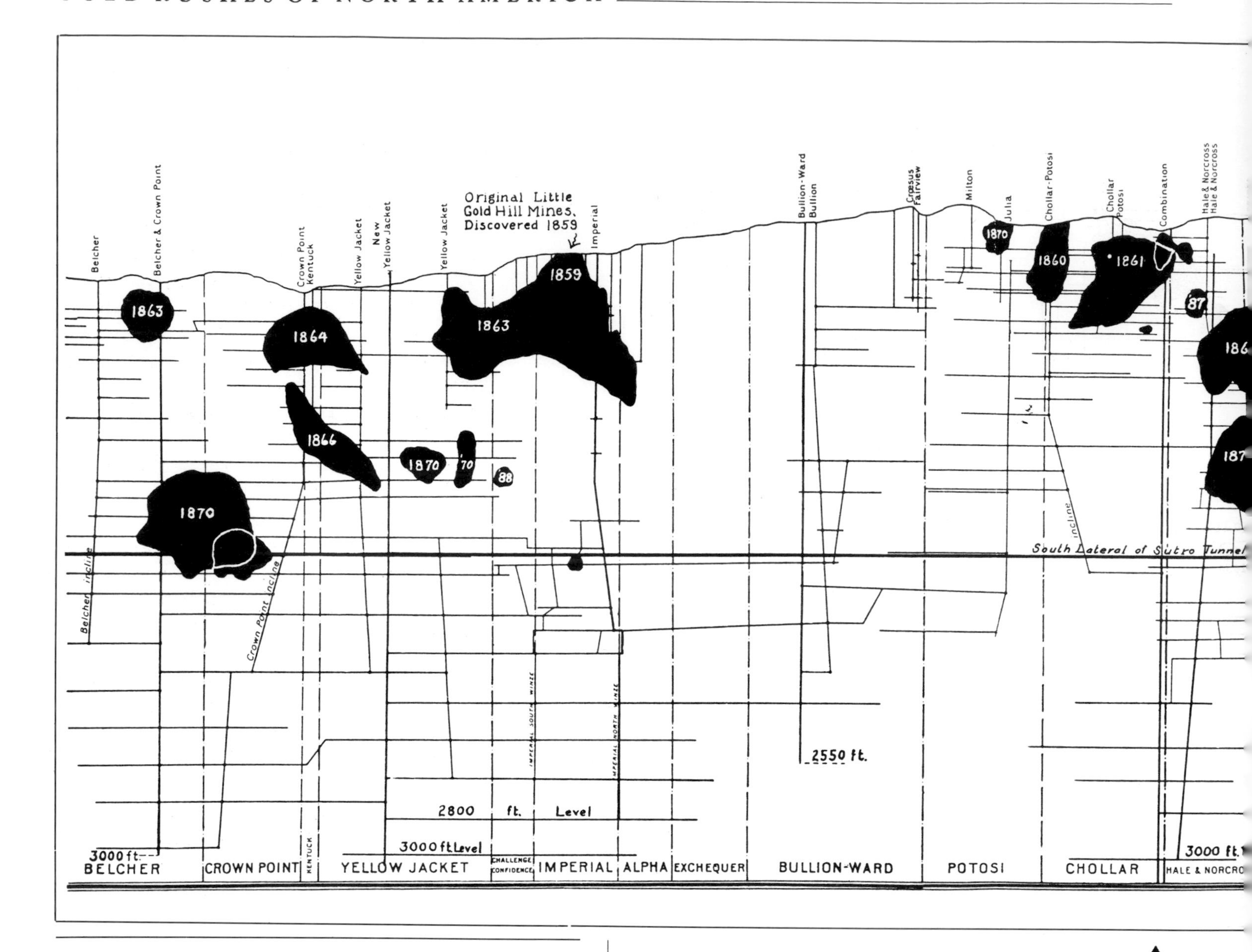

▲ *Vertical section of the Comstock Lode showing principal mines and ore bodies with dates of discovery.*

town of Dayton, Old Virginny attempted to ride a wild mustang while drunk. He was thrown and died a few hours later from a fractured skull. Despite all his drinking he still had $3,000 to his name.

The Comstock Lode has been studied by geologists and miners for years. The odd distribution of the silver and gold ores had a lot to do with the way they were geologically formed. The Comstock Lode was a combination of gold and silver held in suspension by a superheated liquid deposited along irregular fractures in the bedrock formed by subterranean faulting. When the valuable ore laden quartz precipitated out of the solution it did so at varying rates of cooling, piling up in some areas and not others. Sometimes one metal bearing solution under intense pressure squirted past a partially blocked fissure into another whole system of cracks in the rock. This accounted for the uneven pattern of ore chambers, shoots and branches found near Mt. Davidson.

In a historical sense it was found that the Gold Hill placer diggings and the Ophir mine, on opposite sides of Mt. Davidson, were the two ends of the same Comstock Lode that zigzags under and across the eastern face of the mountain. Old Virginny was part of the discovery at both ends of the silver and gold snake. News of this find set off another gold rush. This was not a placer gold rush, but a gold and silver rush. It was a deep shaft mining enterprise that spelled the beginning of the end for the romantic prospector. Labor was plentiful and the salaries cheap. Industrial technology shackled the miner underground. His chains were his wage.

In 1872 a San Francisco newspaper boasted that "Nevada is the child of California." It was a Californian gold rush and San Francisco was its main beneficiary. Nevada residents complained

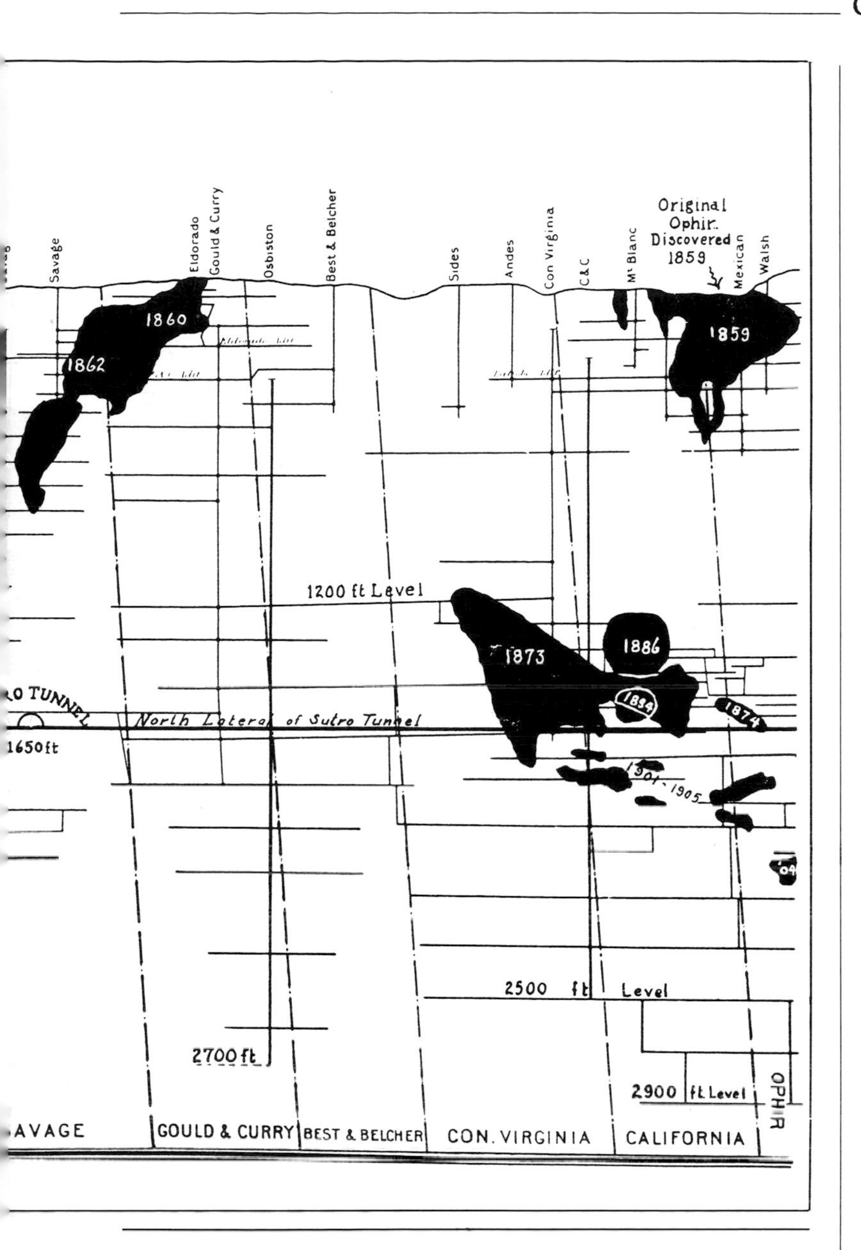

in the 1880s that California stole the Comstock gold and silver and used it to rebuild San Francisco. They said that Virginia City and Gold Hill were run as suburbs of California. And they were right.

When Judge Walsh returned to Nevada City on July 9, 1859, he reported to all who would listen that miners were making between $100 to $400 a day just by using rockers. A man with a pan could average $100 per day. He went on to state that there was little water and timber in the Washoe country and extensive capital investment was needed to develop the Comstock Lode. The roar of men rushing to livery stables for horses and men pushing wheelbarrows full of belongings in the street drowned out the judge's words of caution. In less than a month one third of Nevada City was on their way to the Washoe country to make their fortune at the Comstock Lode.

At best, the primitive dirt trail across the mountains to the Washoe country was wide enough for two wagons. When the road became narrow it was a single lane mountain path with a cliff falling away to a thousand foot drop on one side. On this road came the California gold rushers in wagons, on horseback, in buggies, in stagecoaches or walking with large packs on their backs. These were experienced miners that had a fair idea of what to expect when they reached the diggings. Each man dreamed of being first to the Washoe gold and staking his claim before all the placers were taken. Every day that passed saw an accident claim a life or injure a gold rusher. Soon the road was choked with debris in the form of broken wagons, discarded articles and empty whiskey barrels. But still they came like so much meat being squeezed through a very tight funnel.

Unlike the other gold rushes, once the men reached the Comstock Lode district there was really nothing for them to do. In California, British Columbia and even Colorado, newcomers to the gold rush districts were busy either finding or working claims. In the Washoe country of 1859 all the accessible mines were already discovered and owned. There were no mining jobs available because more capital was needed to start production. Everything was at a standstill. Some men turned around and left while others hung around to await the spring and the influx of new investments. Those who stayed helped to form the greatest Nevada boomtown ever, Virginia City.

According to legend Virginia City was named by Old Virginny. As was his usual habit, he was taking an evening stagger and tripped over a rock. He fell and broke his whiskey bottle. He got up and stared at the shattered glass and the whiskey stain on the ground. Old Virginny had a brainstorm and baptized the spot Virginia. More reliable, if not duller, sources indicate that at the beginning of the Washoe gold rush the town council named the unnamed but burgeoning village Virginia City. The reason for this designation is unknown.

By Argonaut's standards early Virginia City, in the spring of 1860, was a mess. One brief visitor saw the place in biblical proportions: "Imagine a flood in hell, succeeded by a snowstorm." J. Ross Brown, who noted that Virginia City lacked even the most rudimentary planning, painted a more concise picture, "Frame shanties, pitched together as if by accident; tents of canvas, of blankets, of brush, of potato sacks and old shirts; ...smoky hovels of mud and stone; coyote holes in the mountain side forcibly seized and held by men; pits and shafts with smoke issuing from every crevice; piles of goods and rubbish on craggy

Someone at the Ophir mine heard of his reputation and sent for him. It took Deidesheimer one month to invent the revolutionary "square set" method of timbering that is still in use today. It is similar to a sophisticated Tinker-Toy set up. Each beam is mortised and connected to two other thick timbers at each end. Together these timbers form a hollow square, with room to attach another square at either end. To make the whole structure stronger miners would fill the walls of the squares with rocks and timbers from floor to ceiling. After this accomplishment the Comstock Lode was accessible once more.

▲
Miners in Virginia City danced their time away in hurdy gurdy houses while waiting for the Comstock Lode to produce.

It's one thing to get the ore and another thing to process it. In 1861 the richest ore mined at the Comstock Lode was sent to England to be smelted. The poorer ores were dumped, forming large manmade hills near the mines. Smelting is a high temperature melting process that removes silver from the other metals usually associated with it, such as lead, copper or zinc. Almarin B. Paul, a veteran California miner and stamping mill owner, invented the Washoe pan process over

the course of a three year period. His hard work paid off and by 1862 he had revolutionized silver mining. Now all the ores, rich or poor, could be processed with negligible transportation costs and maximum recovery. For 20 years the Comstock Lode was the spark that set off an explosion of technological advances in mining.

Problems of heat as extreme as 130 degrees and found at depths of 3,000 feet were solved by bringing unlimited supplies of ice water into the mines, and having the workers douse themselves during their frequent rest periods. Ventilation was improved by cutting airshafts and installing huge, newly designed blowers. A system of importation of foreign technology and adaptation by Eastern companies was developed. This led to the use of the compressed air drill and the the diamond studded rotary drill by the Comstock Lode mining companies. By 1881 the Comstock Lode was the Western mining show place and the recently created United States Geological Survey commented: "The Comstock Lode is the chief focus of mining activity in the region west of the Rocky Mountains. It represents the most highly organized phase of technical mining which has been reached west of the Mississippi River."

Virginia City went from infancy to early adolescence within two years. In 1862 real money was flowing into the city's economy and with the money came more people. The population grew to over 15,000. One of the newcomers was a former Mississippi River pilot, Samuel L. Clemens. Within a decade the literary world would know him by his pen name Mark Twain. In his first book, *Roughing It*, Twain described living in the the adolescent city. "The mountainside was so steep that the entire town had a slant to it like a roof. Each street was a terrace, and from each to the next street below the descent was forty or fifty feet. The fronts of the houses were level with the streets they faced, but their rear first floors were propped on lofty stilts; a man could stand at a rear first floor window of a C Street house and look down the chimneys of the row of houses below him facing D Street." Everybody and everything was going on in the young Virginia City. Mark Twain saw it all. "There were military companies, fire companies, brass bands, banks, hotels, theaters, "hurdy gurdy houses," wide-open gambling palaces, political pow-wows, civic processions, murders, inquests, riots, a whiskey mill every fifteen steps..." Money may have made the city grow, but it was still a gold and silver rush town.

In keeping with the gold rush tradition, some school teachers came to class armed with pistols. People told a story of the first school teacher in Virginia City who shot a student's ball out of the air during a class break. He was only making a point about keeping order in the classroom. The most popular spectator sport at the time was the bear fights and sometimes, to make the game more exciting, dogs were pitted against the hungry bears in one big free-for-all.

Not all events had this violent edge. On certain evenings miners could be found enjoying Shakespearean plays as well as band concerts and music hall recitals. Once an over the hill singer came to town and gave a concert. She was passed

▲ *In the depths of the Comstock Mines heat wore the miners down more than hard work. Every two hours the miners took a break, drinking and often dousing themselves with ice water.*

her prime and the audience knew it. Instead of booing her off the stage they demanded encore after encore, each time showering her with silver half dollars. By the time the performance was over she had enough money to retire and as her audience had gently but firmly hinted, never performed in public again.

In this rough and tumble mining town was born the legend of the beautiful prostitute with a heart of gold. As with all legends it was a blend of carefully selected facts and conveniently believed fictions. To start with, she was a plain looking woman who came to the Comstock Lode via her birthplace in Liverpool, England. Her point of immigration into the United States was Louisiana, where she married and a short time later deserted her husband. It was here that she took up prostitution as a trade. Soon after she arrived in Virginia City, during the winter of 1861-62, Julia was befriended by her fellow working girls, their pimps and their patrons. In addition to the usual services, Julia offered something the other women could not or would not provide to their customers, companionship.

She dressed well but lived rather plainly. Her small two room cottage lacked a kitchen and indoor plumbing. Julia made money, perhaps enough to live more comfortably, but preferred her cottage instead. It was disclosed at the time of her death that she spent some of her earnings on clothes and jewelry, but most of her money and spare time went to nursing the sick and injured as well as those who were down on their luck. The Volunteer Engine Company No. 1 was also a recipient of her cash.

On January 19, 1867 Julia was brutally strangled and shot in the forehead. She was found the next morning by her part-time Chinese manservant. No one had ever commented on her lifestyle before, but suddenly the whole city was talking. Local newspapers called the killing "The most cruel, outrageous, and revolting murder yet committed in this city." This was an exaggeration because the early days of Virginia City featured several killings each week and it was hearsay in 1860 that "A person isn't respected until he has killed his man." Nevertheless, the newspapers played the crime for all it was worth. "Julia Bulette belonged to that class denominated 'frail but fair,' yet being of a very kind-hearted, liberal, benevolent and charitable disposition. Few of her class had more true friends." It was decided by her friends and the press that Julia's "faults were to be buried with her and her virtues were to live on."

A short time later John Milleain was found to have several items of Julia's in his trunk. The police were alerted by a jeweler to whom Milleain had attempted to sell a diamond pin. It was the same jeweler who had sold the pin to Julia. It was a quick trial and Milleain claimed he was framed right up until the time they hung him in front of 4,000 people.

By 1869 Virginia City had endured a great fire in the business district, the rebuilding of the business district, a spasm of Vigilante committees and the growth period for schools, churches, city officers and police. Things were settling down.

During the early years the earth under the city was honeycombed with tunnels and some of the miners worked in the very tunnels that ran under their own houses. The quality of life for all the miners took on the dust of industrial dreariness. At the beginning of each shift the miners were collected in a large room. The timekeeper called the roll and checked each man again as he passed the gate leading to the mine. The same double check was made for those men leaving at the end of their shifts. By the mid-1860s large absentee owned companies were the rule. Even by the anti-trust standards of the day these companies exercised the powers of a monopoly.

From the miner's point of view, many of the technological improvements made the job easier but work in the mines more hazardous. The air drill that made the process of creating holes for explosives simpler also released volumes of lung-clogging dust, which in turn lead to black lung disease or silicosis. The powerful new explosive, dynamite, which was introduced to the Comstock Lode in 1868, increased the number of accidents due to premature or delayed explosions. With the introduction of the braided steal cable, mine elevators came into widespread use. Occasionally, workers who were fatigued from working in the hot, fetid air of the mines and unfamiliar with the new moving platforms and cages suffered attacks of dizziness in the elevators. Every once in a while a miner would fall between the cage and wall suffering injury or death. At the Comstock Lode the free-spirted gold rusher was an anachronistic memory and industry was on the march.

In 1872 the great wealth of the Comstock Lode appeared to be over. Then in 1873 the Big Bonanza was found at 1,167 feet. Much of the money from this incredible strike found its way into the hands of only a few men. Four Irishmen, James Fair, James Flood, Jack O'Brien and John Mackay owned the mines in which the Big Bonanza was found. Despite their wealth, the Bonanza Kings were the most conscientious and honest men in the mining business during that day and age. All four were "hands on owners" who went into the mines to work daily. Mackay, a former carpenter, could look at sample of ore and discern its silver content within a half of a percentage point of the official assay. He and O'Brien arrived in the Washoe country from California in 1859. As they gazed at Mt. Davidson O'Brien and Mackay made a pact. They vowed to begin their mining ventures "...Like gentlemen" and threw away their last silver half dollar. At the time of his death in 1902 his business manager was quoted as saying: "I don't suppose he knew within 20 millions what he was worth."

In 1873, the year of the Big Bonanza, $21,000,000 was taken out of the mines. The high point came in 1878 when the total gold and silver mined from the Comstock Lode came to $38,000,000. San Francisco was the the destination of all this wealth and in a short time millionaires were jumping up like popcorn. Four years later the Bonanza Kings found that the rich ore was running out. Millions of dollars were spent to drive the tunnels deeper, from 1,650 feet to 3,000 feet. At the same time miners not working on the speculative tunnels were laid off and the exodus of unemployed miners from Virginia City began. Tragedy struck in 1882 when an immense reservoir of hot water was pierced and subsequently flooded the deep level mine shafts. The miners were forced to return to the upper levels and work on the less productive ores. The glory days of the Comstock Lode were over.

Much has been said about the silver mined at the Comstock Lode. Silver was the major ore mined, but gold started the rush and the percentage of gold was second by a small margin. From 1861 to 1881 the Comstock Lode's yield in silver was 57 percent, compared to the 43 percent yield in gold. It was said that during its boom years the Comstock Lode made six men enormously wealthy, about 30 men millionaires and gave a large number of men estates worth about $100,000. The wealth was distributed disproportionately to the number of people who participated in the mining of the Comstock Lode. George Hearst and other early investors began their fortunes at the Comstock Lode, but they still had to work for their subsequently incredible fortunes. Later, men such as the four Bonanza Kings had their bounty literally dumped in their laps.

The profits from the Comstock Lode quickly migrated to California, particularly San Francisco. For San Francisco the bonanzas were especially fortuitous. It allowed the city to finance new buildings, some of them quite ornate, as well as diversify its businesses. San Francisco used the Comstock Lode based prosperity to solidify its position as the leading city on the Pacific Coast.

The Comstock Lode truly ushered in the era of industrialized mining. Technology was the key to extracting the rich ores buried under Virginia City and Mt. Davidson. With the advent of these new industrial mining methods came a change in the status of the miner. There would still be more gold rushes, but the die was cast and the placer miner was to face extinction as the buffalo did on the plains.

CHAPTER 9

CHASING THE GOLDEN DRAGON

Until the Klondike and Yukon gold rushes at the end of the 19th century, there were a series of minor, but still great, gold rushes in Idaho, Montana, the Black Hills of Dakota and few brief spasmodic rushes in Arizona, New Mexico and Utah. All these rushes had the quality of a drug addict chasing his first high. The drug addict never quite recaptures the initial feeling, the exhilarating surprise of the virginal mind numbing rush, yet he keeps consuming more and more drugs in the attempt to recapture the moment. The Chinese first observed this among their own opium addicts and called this fruitless, obsessive behavior "chasing the dragon." Here are few representative examples of chasing the golden dragon.

IDAHO AND MONTANA

Except for a seacoast, almost all the geological features typical of the continental United States can be found in Idaho. It has jagged mountains and broad valleys, green prairies and arid plains, humid forests and sagebrush deserts, prehistoric limestone coastlines and volcanic deposits. Sprinkled among the broad features are canyons, waterfalls, rivers, streams, lakes, hot springs, sand dunes, hills and glaciated peaks. All these features are contained in an area practically twice the size of New England. Idaho is the 13th largest state.

Originally part of the Oregon Territory, Idaho had been claimed in turn by Spain, Russia, Great Britain and the United States. The last two claimants settled the matter with the Oregon boundary settlement of 1846. The basis for America's claim to Idaho was the explorations of Lewis and Clark between 1805 and 1806. These two intrepid explorers noted that the area was populated by diverse groups of semi-nomadic Indians. This nomadic lifestyle was aided with horses traded to the Indians by small bands of Spaniards who traveled through the land a century earlier. From 1808 until the 1840s Idaho was the center of some fierce competition among several fur trading companies. Beaver, which was used in making hats, as well as the always popular mink and ermine weasels, were the early 19th century craze. In the 1840s overexploitation as well as a style shift from fur to silk hats reduced the large fur trading corporations to smaller operations. Idaho had to wait for the Clearwater and Salmon River gold rushes before it attracted permanent settlements.

The year was 1859 and the Indian wars in the interior of the Washington Territory, originally part of the Oregon Territory, were over. During the several years of conflict the United States Army navigated the Columbia River and used it as the main water highway into the interior. Now those who believed there was gold in the territory flocked to the interior using the newly charted river. It was among these prospectors that rumors circulated concerning the possibility that Snake River, a tributary of the Columbia River, might be gold bearing. Elias D. Pierce, a trader with the Indians, decided to investigate these stories in 1860. He found gold near the southern branch of the Clearwater River. The main problem with his find was the fact that the gold was inconveniently located on Indian lands. Pierce had to evade the Indians and Army patrols to return, with his samples, to his base in Walla Walla. When the hoard of miners showed up the next spring there was nothing the Army could do to stop them, so they wisely negotiated a treaty with the Nez Percé Indians. The U.S. government negotiated a treaty in 1863 with the Nez Percé that promised to protect the reservation if the Indians would allow a site for the permanent town of Lewiston to be built on part of their land. There was also a provision for some monetary compensation. But not all of the Nez Percé agreed with the treaty and a small band led by Joseph, father of the famous Chief Joseph, stayed in the Wallowa Valley where Lewiston was situated. At the same time a volunteer company of miners marched through Idaho in a show of strength. When the Indians got upset and the miners got trigger happy. Some Indians were wounded or killed and there was no further trouble.

It was a diffuse gold rush with miners rushing all over the place. West Coast historian Hubert Howe Bancroft described the prospector's activity in his book, *History of Washington, Idaho, and Montana*: " The miners of Idaho were like quicksilver. And many of them dropped in any locality broke up into individual globules and ran off after any atom of gold in their vicinity. They stay nowhere longer than the gold attracted them." They behaved in this manner because they were placer miners, and by and large the placer deposits ran out quickly.

While the placer miners were averaging about $10 per day, the more organized miners quickly set up flumes and sluices and quadrupled the amount of gold recovered. This meant that many more miners found themselves in the position to work on claims other than their own and make more money.

For those who still dreamed of striking it rich the more inaccessible country beckoned. About 1,000 prospectors moved into the Salmon River country. This was not the rugged canyon of Fraser or one of the out of the way valleys of California, but a marshy plateau 6,000 feet above sea level covered with black pines and tamarack trees. It was tough going with no trails to follow and wet muddy ground to traverse. Even in the winter when the only food that could be brought in was flour, new miners came into the country on snowshoes. These men were determined to work for themselves, be their own bosses and perhaps strike it rich in the process. They did not know they were independent cogs rotating around the growing wheel of progress and that wheel would leave no room for the little cogs in the future.

On March 4, 1863 the Idaho Territory was created to include all the new mines in Idaho and most of Wyoming and Montana. The final boundaries of the future states would not be set until 1868. More than 32,000 miners and merchants, packers and farmers had moved into the newly organized territory. The gold rush had settled over 16,000 of new arrivals in the Boise area. About a third of the immigrants were in Montana. The rest were scattered, in small numbers, throughout the large territory.

Montana is the fourth largest state, after Alaska, Texas and California. Because of its rich mineral resources it is called the Treasure State. The eastern 60 percent of Montana is in the Great Plains which contains many small streams and rivers. A visitor's first impression of the Plains is one of lushness. This is due to the abundance of water, trees and grasses. Abruptly shooting up from the western part of the Plains are isolated mountains. They stand like lone sentries guarding the Rocky Mountains. The Rocky Mountains with their sharp peaks and valleys occupy the extreme western 40 percent of the state. Montana lies directly east of Idaho.

Until Lewis and Clark's expedition through the area the only people, other than the Native Americans, who were living in the state were French traders and fur trappers from Canada. Following the same trend as in Idaho, fur traders were the largest group of immigrants to the area until 1860 when James and Granville Stuart publicized their gold strike on Gold Creek near the Deer Lodge district.

The first person to discover gold in Montana was François Findley. He regularly traded trinkets and cloth with the Indians in exchange for buffalo hides and furs. On one such journey, in 1852, he was struck by the physical similarities between the gold country of California and the river and bar formations of Benetsee Creek, near the Deer Lodge area. Findley used his metal campfire pan to sample the sand and gravel of the creek. To his amazement he actually panned a small amount of gold dust. But Findley was not a man given to rash actions or excited emotions. In reporting his find to his supervisor, Angus McDonald, chief factor of the Hudson Bay Company at Flathead Lake, he was restrained in his presentation. Nevertheless, McDonald advanced the food and supplies necessary for a further investigation of the area. Findley was definitely not a prospector; he was lucky the first time. Findley did not find any more gold in the Deer Lodge district and many at the Hudson Bay Company gave a sigh of relief. The last thing they wanted was thousands of miners swarming over the country, panning for gold and destroying their fur trade.

Six years later James and Granville Stuart, along with some friends, were returning to Iowa from a not too successful trip to the California gold fields. The party stopped in Deer Lodge Valley for the night. In the morning the group panned some gravel, having noticed the same similarities in the countryside that Findley did. Their reward was some small specks of gold dust at the bottom of each pan, worth about 10¢ at the going 1858 rates. They endured tasteless meat, poor equipment and stolen horses because they believed there was gold in the area. Finally they left without finding the big strike. Two years later they returned a little wiser and better prepared. This time they found rich placer deposits, so much so that James and Granville wrote to their other brother Thomas, who was prospecting in Colorado, and asked him to partake in their rich find. Thomas, unlike Findley, was not reserved, and he told a few of his close associates about the find. Thomas immediately left Colorado for Montana with about 15 of his friends.

By 1862 the Stuarts had installed the first system of sluices in Montana and were mining gold like crazy. They did, however, miss a much richer deposit on the grassy hill above the stream where they kept their horses penned in. But all they knew about was creek mining, and the hillside gold had to wait four more years before being discovered by other miners. As the rush gathered momentum new sites were discovered. Grasshopper Creek, discovered by John White in 1862, was a particularly rich find. So fast was the development near Grasshopper Creek that a large mule train left for Salt Lake to gather supplies in the fall of 1862. Upon its return in the spring of 1863, it found the town of Bannack laid out, built and populated by 800 miners.

In spite of the richness of various finds in both Idaho and Montana, few roads were built. Idaho with its mountainous western portion was the more accessible due to its proximity to California and the Columbia River. Traveling through Montana's plains without roads was a nightmare for 19th century miners. Wagon wheels got stuck, there wasn't enough game for the travelers to hunt and there was the constant threat of Indian attacks. In any case transportation within the territories and to the outside world was difficult and expensive. Added to transportation hardships were the severe winters. Mining had to be done at a fast and furious rate in order to earn enough in eight months to support the gold rusher for the whole year. These geographical and climatic factors alone could make life difficult enough for the gold rushers. But there were also historical forces to be reckoned with.

Because of its gold and remoteness, the Idaho Territory attracted men who for one reason or another had lost their livelihood on account of the Civil War, along with draft dodgers, deserters, displaced guerrillas and Confederate sympathizers. Many of the latter group had only one trade, violence, and they plied it whenever the need for money arose, which was quite often. Since these were mining camps and towns it was easy for men to account for sudden wealth. Retribution was scarcely a worry; law enforcement was an unknown quantity in the unsettled land.

Many of the early desperados were men such as David Howard, D.C. Lowry and James Romaine. In August of 1863 they plotted to rob a rich trader and pack train owner named Lloyd Magruder. The trader had just left Lewiston and was on his to sell his wares in Virginia City, Montana. Howard, Lowry and Romaine followed the mule train. Before they caught up with Magruder they were joined by a trapper, William Page. The trapper knew nothing of the party's intentions toward Magruder; he was going to Virginia City to sell his furs. When the gang joined Margruder's pack train they became the most helpful traveling companions a trader could journey with. No favor was too big for the conspirators. When they reached Virginia City, Howard, Lowry and Romaine helped the trader sell his goods. They also kept close tabs on how much money Magruder had made. As far as they could tell the trader had over $25,000 on his person when he began his return to Lewiston. Much to the outlaws' dismay Magruder also picked up four other men from Missouri who were also bound for Idaho.

Somewhere at the halfway mark the three conspirators took Page aside and told him they were going to kill the rest of the party that night. He was given the choice of either joining them and sharing in the loot or dying, he choose to save his own life and make some money in the bargain.

That night Magruder and his four innocent traveling companions were hacked to death with an axe. Howard, Lowry, Romaine and Page kept one horse and seven mules. They threw the bodies along with the remaining mules off of a cliff.

The story should have ended here with no one ever knowing what became of Magruder. But something very strange had occurred in Lewiston the same night. Hill Beachy, a hotel owner and friend of Magruder, had a dream. He dreamt that his friend and four men he did not know were axed to death by some desperados, three of which were vaguely familiar to him. Beachy awoke in a cold sweat, told himself he was having a nightmare and fell back into a fitful sleep. By the next morning he all but forgot about his dream.

By all rights the four outlaws never should have gone to Lewiston. They planned to avoid the town, but when they attempted to cross the Clearwater River above the town they found it flooded due to unusually heavy summer rains. So the bandits sneaked into Lewiston after leaving their horses with a rancher outside of town. Page booked passage for the group on the next stage to Walla Walla. The natural place to do this was at a hotel, and the hotel owner who took the reservations was none other than Beachy. Beachy took one look at Page, noted his shifty manner and became suspicious. He looked like as if he had something to hide and Beachy was compelled to find out what it was.

Following Page was easy. Beachy knew the town and how to keep out of sight. Page sneaked back to his new found friends sitting in the back of the saloon. It took just one look for the hotel owner to recognize the badly disguised Howard, Lowry and Romaine as the three shifty men who had left town shortly after Magruder. A few seconds later the dream of Magruder's death welled up like gas bubbles from a murky lake floor into Beachy's consciousness. He was so stunned by the vision that he didn't see the quartet leave the saloon.

Howard, Lowry, Romaine and Page were already on the stage to Walla Walla, but Beachy continued his investigation. He traced the outlaws' abandoned horses to a nearby ranch. On one of the horses was Magruder's saddle. Beachy knew at once what had happened to his friend and that Howard, Lowry, Romaine and Page were guilty of Magruder's murder.

Once more the story should have ended with the discovery of the saddle but Beachy was a man with a mission. He preformed the impossible task of having warrants sworn out for the quartet's arrest. Next, he had himself appointed deputy sheriff and for good measure had extradition papers drawn up for California, Oregon and Washington. Beachy was not going to risk any foul-up when he caught up with the men.

Tracking the killers turned out to be easier than Beachy expected. He found them at the San Francisco mint waiting for the gold dust they stole from Magruder to be made into gold coins. In December they were extradited back to Lewiston, and by this time the bodies had been found with Page's help. Page turned state's evidence and received a short jail sentence while Howard, Lowry and Romaine were hung.

But perhaps the most notorious character to ever terrorize the Idaho Territory was a 24-year-old Connecticut Yankee named Henry Plummer. He fled from his past to the remote Idaho Territory in 1861. When Henry Plummer arrived in Lewiston he was considered by the townspeople to be a wealthy, energetic, civic minded, New Englander who came to town with his wife. Plummer quickly became a leading member of the community. What the unsuspecting folks of Lewiston did not know was that he would eventually make their lives miserable.

His career started out respectably enough when he emigrated to Nevada City, California in 1852 and started the Empire Bakery. Soon the popular young baker was elected marshall of the town. He was not marshall for long before he went to prison for shooting an irate husband whose wife was having an affair with Plummer. But he didn't spend a long time in prison either. His friends had access to the governor and Plummer was granted a pardon for the bogus reason of being terminally ill from tuberculosis. He killed his next victim in a brothel soon after he was released. Now Plummer was on the run and needed money. In the Washoe Country the young outlaw attempted to rob a Wells Fargo stagecoach, but the robbery was foiled when the pistol Plummer was using failed to fire. The driver escaped with his life and never showed up at Plummer's trial, so the the young outlaw was once again set free.

It was this last escape from imprisonment that sent Plummer into the partially isolated Idaho Territory, but not before he killed another man in quarrel over a woman, was arrested for the crime, bribed his way out of jail, stabbed the sheriff who arrested him, fled toward Oregon while spreading rumors of his death to prevent pursuit, stole another man's wife and horse in Walla Walla and finally made his way to Lewiston.

In the beginning it was a case of camouflage. There were so many desperados in and out of town that Plummer, whose official occupation was that of a gambler, had the image of being a rather honest man, given the local circumstances. Soon a new phenomenon occurred around the area of Lewiston. There appeared a group of highwaymen, called road agents in the Idaho Territory, who would rob anyone, miners in particular, of their money, gold, horses and sometimes even their lives. Although there were those few who suspected that Plummer was the secret boss of the road agents, they kept silent for fear of being murdered in the middle of night. When the truly naive protested the wave of terror Plummer would appear to be the voice of reason and head off the formation of any Vigilante groups.

In less than a year there were 21 murders committed by the road agents, and in the fall of 1861 there were four murders in the town of Lewiston itself. It was so bad in Lewiston that having a horse stolen at gun point, in broad daylight and in front of a dozen witnesses was considered to be just an ordinary event and nothing to get excited about. Even when a popular saloon keeper was viciously murdered in his establishment for protesting the wave of terror, the townspeople behaved as obedient sheep and went about there daily business. To make his growing criminal empire more efficient, Plummer set up two roadhouses, or "shebangs" as they were called, for his henchmen. These roadside way stations were located near the trails most frequently used by travelers. They were employed to collect whatever fees they could from unsuspecting travelers.

Plummer decided to expand his operations to include Florence. As before he kept a low criminal profile and acted as a spy, telling his men who was carrying how much gold dust out of town and where they were heading. In October 1862 pack train operators John and Joseph Berry left Florence after delivering goods to the town. They were on their way to Lewiston. Plummer sent word to his gang that the brothers were carrying $2,000 in cash. About 50 miles away from Lewiston the Berrys were robbed at gun point. There was nothing unusual about the hold-up. Following the usual script the hold-up men wore masks and threatened to kill the brothers if they didn't give them the money, but the Berrys recognized two of the bandits by their voices. After the robbery the brothers followed one bandit to a saloon in Walla Walla where they arrested him at gunpoint. The other outlaw was found and taken into custody at a saloon in Wallula. They were extradited to Lewiston and remained nonchalant about their arrests. They expected to be rescued by their cohorts within a day or so.

The pendulum of justice was beginning to swing toward the side of law and order in Lewiston. The word out on the dusty streets was that any suspicious characters seen in town would be arrested on sight, and anyone who resisted would be shot. When faced with a determined citizenry, Plummer's gang left town and faded into the surrounding countryside.

A new district judge was appointed and he rode to Lewiston from Olympia. Plummer secretly paid for an attorney to defend his two gang members. The clandestine gang leader thought he would free his men by a crooked trial and put an end to this twinge of decency that was infecting the townsfolk. He was wrong. When the lawyer arrived at the jail to consult with his clients the jailors told the attorney that visiting hours were only in the morning and he had to wait until the next day. The next morning when the lawyer arrived a second time he found the jail unguarded. He cautiously went inside and found his clients hanging from the ceiling beams doing what the miners called "a mid-air dance."

Plummer didn't survive all this time by being stupid. He left town in hurry for Bannack, Montana. By now the whole Snake River country was full of angry citizens just looking for bandits to hang, and the outlaws found that their former stock and trade was becoming a dangerous enterprise.

Things were going well for James and Granville Stuart. Their system of sluices was producing gold and they had just finished building their ranch complex at Deer Lodge. On the night of September 16, 1862 the brothers had a very charming visitor who was a gambler from the West, Henry Plummer. Much to Plummer's surprise he met one of his desperado friends, Jack Cleveland, at the Stuart ranch. Cleveland had fled Idaho at the same time and for the same reasons. Together, both outlaws fell madly in love with the same woman, Eliza Bryan. They rented a cabin, divided up the chores and wooed the fair young lady. Plummer won, of course, and married Eliza a few months later. The newlyweds moved to Bannack, leaving a a very rejected Cleveland to sulk.

Right after New Year's Day Plummer made his presence felt in Bannack. Most of the townspeople believed him to be a urbane, pleasant kind of guy, the type of fellow who should be elected to a city position. However, during those other hours when the honest folks were busy, asleep or just not paying attention to Plummer's comings and goings, he was building another outlaw gang of 100 strong. Unknown to the townspeople,

Plummer bought a small spread and called it Rattlesnake Ranch. In his book, *The Vigilantes of Montana*, Thomas Dimsdale followed Plummer's career of crime and gave his readers some idea of what the place was like. "Two rods in from this building was a sign post, at which they used to practice with their revolvers. They were all capital shots. Plummer was the quickest hand with his revolver of any man in the mountains. He could draw a pistol and discharge the five loads in three seconds." It was outlaw heaven.

The dark cloud on Plummer's horizon was Cleveland. He was the only man within hundreds miles who knew Plummer's true nature, and unwittingly Cleveland provided the solution to Plummer's dilemma. One fine evening Cleveland, who was said to have murdered several miners, walked into a saloon and accused a man named Perkins of failing to pay up a loan. Perkins denied the allegation stating that he had already paid him. But Cleveland insisted he pay up the loan and drew his gun. It had been Cleveland's experience that pointing a gun at someone always made him yield his money, but Perkins just stood their glowering at Cleveland. At this moment Plummer stepped in and stopped the fight. He told Cleveland to behave himself. Something snapped in Cleveland and in a torrent of profanity he threatened to kill both Perkins and Plummer. "This has gone far enough," Plummer said as he drew his pistol and shot the startled Cleveland in an area slightly below his groin. The wounded man fell to his knees grasping his wound and crying, "You would shoot a man when he is down?" "No, get up," said Plummer, and when Cleveland managed to feebly get to his feet his former friend shot him two more times.

Cleveland was taken to his rented room by the sheriff and his deputy. When asked what the fight was about he told the sheriff that it was none of his business. Two hours later he died. Plummer thought Cleveland had said something to the sheriff and deputy that might reveal his true nature. He vowed to get rid of them, but first he had to stand trial for murder.

The trial was a short one. All the witnesses, who detested the dead man, agreed that Cleveland had threatened to shoot Plummer and that Plummer, the peacemaker, had no other choice but to shoot the scoundrel in self-defense. Plummer was acquitted and soon afterward the sheriff was run out of the territory by Plummer, but not before shooting Plummer in his right arm. From that point on the bandit chief had to draw his pistol with his left hand. The townspeople needed a new sheriff and the most popular candidate was Plummer. Plummer's election signaled the beginning of a new wave of lawlessness in Montana.

His gang called themselves the "Innocents." By now the outlaw band was so large that when confronted by a potentially honest person they would utter the code phrase, "I am innocent," and often that was enough to identify one gang member to another. Plummer was a genius of crime and if he had lived in London he would have been the prototype for Sherlock Holmes' nemesis, Dr. Moriarity. His spies were everywhere and nothing of value was safe. Sometimes Plummer would ride out of Bannack on "official business" only to don a disguise and ride with the Innocents as they robbed another stagecoach, pack-train or whatever. When he returned those who had their suspicions said nothing, because that was the only way to survive in Montana at the time.

Gold was discovered at Alder Gulch, Montana, and from the sites came enough money to build Nevada and Virginia City. Mingled with the gold rushers flocking to the diggings were small groups of Innocents. Once more the pattern repeated itself and the miners' gold fell into the waiting hands of Plummer's criminal empire. As he collected the money, Plummer would spend it to further his honest image. On Thanksgiving Day he held a large dinner and invited all the prominent members of Montana society. The first turkeys ever seen in Bannack were on his table, imported from Salt Lake City for the sum of $40 apiece. Unbeknownst to Plummer were the few at his table that Thanksgiving Day who had a very clear idea of who he really was and what he had been doing. Events had been catching up to him, but he believed in his own invincibility and chose not to notice.

One of the first to suspect that something was wrong with Plummer's honest image was his next door neighbor, Col. Wilbur F. Sanders. Plummer had told the colonel that he had discovered a possible silver site and had to leave town for a couple of days. This was his excuse to take part in a stagecoach hold-up. Sanders was always on the lookout for gold and silver deposits wanted in on the claims filing. As the sheriff of Bannack was leaving to rob the stagecoach he was approached by Sanders. Sanders told Plummer that he was going to travel with him to the silver site. Plummer was taken aback and mumbled something about going to Rattlesnake Ranch first and

▶
A miner working the rocker in Alder Gulch, Montana. Lucky was the gold rusher who worked Alder Gulch and did not become a victim of Plummer's gang.

rode off. Undaunted Sanders trailed Plummer, keeping an hour behind the sheriff. When Sanders arrived at the ranch he inquired about Plummer, who was now on his way to rob the stagecoach, and was told he was unavailable.

Sanders spent the night on the floor of the empty bunkhouse. That night a drunken friend of Plummer's came into the bunkhouse, didn't recognize Sanders and tried to shoot the colonel. Sanders was quicker than the drunken outlaw and took his gun away. At this point the man recognized Sanders, apologized and passed out. Throughout the night the outlaws returned from the hold-up talking about the robbery. No one noticed that there was a stranger in their midst because there were always new faces hanging around in those days. Pretending to be asleep, Sanders heard them talking and learned more than he wanted to know. That morning he rode home thankful for not being discovered.

The next person to uncover the truth about Plummer's double-life was Henry Tilden, a young employee of Chief Justice Sidney Edgerton of the Idaho Supreme Court. Tilden was on a horse drive when several masked rustlers rode up to him and robbed him of the horses and his money. A sudden gust of wind blew a bandit's mask aside. Although Plummer did not know the young man, Tilden immediately recognized Plummer. The hold-up victim kept his wits about him, showing no indication that he had seen anything. When the outlaws released him, he rode straight to Edgerton and told the judge what happened. At first Edgerton refused to believe the young man's crazy story, but Tilden's earnest presentation won the judge over. Soon there was a small but dedicated group of influential men planning to rid Montana of the criminal.

One of the most notorious of the Innocents was George Ives, a tall, strong, blond haired, amoral man with a very sadistic sense of humor. He too shared Plummer's fondness for shooting men around the area of the groin. He loved to joke with his victims, making them sweat it out before killing, maiming or letting them go unharmed. Ives brutally murdered a young German, Nicholas Thiebalt, whom everyone thought was Dutch.

Thiebalt went to Ives ranch at the request of his employer, Henry Clark, to pick up some mules. The ranch served as both a boarding range for other ranchers and miners' animals and a secret meeting place for the local band of Innocents. While Thiebalt was paying for the mules keep Ives and his band noticed that the young man was carrying over $300 in gold dust. The Innocents knew good mules when they saw them and were reluctant to let this opportunity pass.

When Thiebalt did not return with the mules Clark thought he had been too trusting of the young man and now had to pay the price. But a month later Thiebalt's body was found. The young man had been dragged along the ground until his face was almost worn away and then shot in the head. After taking the body to Nevada City, the furious Clark vowed to avenge the murder. Part of Clark's plan was to arouse the righteous indignation of the town's citizens, so he had the body put on display for a couple of days. Some of the people were shocked, while other attitudes can be best summed up by a miner who said, "People are killed and buried in town every day and nobody asks any questions."

Twenty five men met a few nights later and formed a Vigilante committee, with Clark as its chief. They had a good idea who killed Thiebalt and rode out to Ives' ranch to confront the killer. As luck would have it they first grabbed an Innocent who had taken part in the killing. When the outlaw saw that Vigilantes meant business he quickly turned against Ives. Ives was arrested and after two escapes he was brought to trial in Nevada City. The remorseless outlaw demanded to be tried in Virginia City because he would then be assured of escaping with the help of Plummer's gang. It was not to be. The Vigilantes were not taking any chances with the monster. As soon as Ives was found guilty he was hung. This event gave heart to the fearful folk in Montana and many of the sheep became lions. The day after Ives was hung another Vigilante committee sprang up in Virginia City, and soon afterward small determined committees sprouted like mushrooms across Montana.

Still no one could prove anything against Plummer and the Vigilantes would not move until they had something substantial. This evidence came in the form of a confession from an Innocent, Red Yager. He believed that by telling the Vigilantes how the gang worked and who the mastermind was he would save his own neck. Yager was found one morning hanging from a tree with sign around his neck, "Red, Road Agent and Messenger."

Plummer did not believe that he was vulnerable. The once astute and alert man was drunk with power. His end came on the icy cold night of

◀
After Plummer's gang was hunted down and at last forced to leave the territory, miners could safely return to the business of prospecting without fear of being robbed, as did this man panning near Virginia City, Montana.

nesses. Concurrent with the arrival of the fur traders was the arrival of the Sioux Indians. The Sioux called themselves the Dakota. It was the French who coined the name Sioux after the Algonquin term meaning enemies. The newcomers, having easily adapted to the horse, came out of the forests of Minnesota and pushed the Arikara, an agricultural and village dwelling people, north.

Under Napoleon, France reclaimed the area as part of the Louisiana Territory. It was under French rule for only three years before being sold to the United States as part of the Louisiana Purchase. Unlike other gold rush areas, South Dakota is well east of the geographical area that extends from the Rockies to the Pacific, and settlement of the area started before there was even a gold rush.

Much of the early settlement was instigated by land speculators from Iowa and Minnesota who were interested in the abundance of water power, especially at the falls of the Big Sioux River. Several treaties with the Sioux made some of these lands available to the incoming Americans. A permanent settlement was begun at Yankton in 1859, and three years later when the Dakotas became a territory Yankton was named as its capital. During the decade of the 60s the territory's population fluctuated with the outbreak of the Indian Wars in 1862 and the uneasy peace that followed. In all, about 12,000 people migrated to the area by 1872 when the first railroad connecting Sioux City and Yankton was completed. After this the population rose steadily, partially because of the gold rush in the Black Hills.

The Black Hills, located in the southwest corner of South Dakota, were so named for the somber green evergreens that appeared to be jet black when seen from a far distance. At some unknown time the Sioux discovered gold in the Black Hills and occasionally they showed their finds to traders. The traders never told anyone about the gold because it was not their trade and they did not want to see the land overrun with gold rushers. The noted Jesuit Missionary Father DeSmet, who migrated to the area in 1848, had close ties with the Sioux Indians. When he was shown samples of gold taken from the Black Hills DeSmet told the Sioux to guard their secret well, or else the ever greedy white man would destroy them.

In order to understand the events of the 1870s it is necessary to understand the Sioux and their wars with the encroaching white man. When the Sioux were first introduced to the horse it was a chance to realize their dreams. The Plains had a mystical pull on the Sioux and the horse gave them the means to conquer the wide expanse of seemingly endless land and sky. The horse gave rise to incredible feats of riding, ambush, fast war tactics and warrior nobility. Some believe it was the Plains themselves that brought about these changes. One observer noted that when a man rides in a country that seems to go on forever, he feels closer to God or sometimes like a god himself.

In 1865 the federal government started construction on the Bozeman Trail, a short cut to the Montana gold fields that ran from Fort Laramie to Bannack. It cut through the Indian's main hunting land in Wyoming, and the Sioux chief, Red Cloud, saw this as reason enough for a war. In all fairness the Sioux were asked to sign a treaty ceding the land to the U.S. government, but Red Cloud refused and the Army built forts along the trail to protect the supply line.

The overall strategy was simple, through a series of ambushes the Sioux and their allies, the Cheyennes, would close the trail. No one, save a large armed force, wanted to risk the route for fear they would be attacked. At the same time they besieged Fort Phil Kearny, at the foot of the Big Horn Mountains. After close to a year of constant, stinging attacks that caused the fort to be evacuated, Capt. William J. Fetterman asked his commanding officer to, "Give me 80 men and I'll ride through the whole Sioux nation." It is said that the worse thing that could happen to someone is to have his wishes granted. In Fetterman's case this was certainly true.

In December 1866 the hot headed captain, not known for his knowledge or love of the Indians, was given 81 men to punish the Sioux for attacking a wood detail. He charged out of the fort and after a group of 10 Indians, including the young Crazy Horse. The small Indian band led Fetterman on what seemed to be a wild goose chase until the troop's horses tired. Then, without the means of a swift retreat, they found that they had been led into a fatal ambush and were wiped out. During the following 18 months the U.S. Army was able to inflict two small defeats on the Sioux and there allies. These victories were the result of the Army's new breach loading repeating rifles and not due to any change in tactics by the besieged troops. In 1868 the U.S. government called it quits. They left their forts, abandoned the Bozeman Trail and reserved the Powder River

▶ *Chief Joseph, leader of the Nez Percé, led his people on a long retreat to Canada in a campaign that made military history.*

country and Black Hills forever. Red Cloud and his warriors had the satisfaction of burning Fort Phil Kearny to ground. But forever was not as long as everyone thought it would be.

Enter Lt. Col. George Armstrong Custer. Graduating in 1861, he finished last in his class at West Point. In a rush to build up the Army at the outbreak of the Civil War, his West Point standing and a minor court martial were overlooked, which was a good thing for the Union Army. His daring and brilliance in battle led to a series of promotions culminating, at the age of 24, with the rank of major general in 1864. When the war ended and the volunteer army disbanded, he was forced to revert back to the rank of captain for a few months after which he was given the rank of lieutenant colonel. He was already a hero when he led the 7th Cavalry in the Indian Wars of the Plains. Custer became known as a fierce Indian fighter because of those battles and was praised by commanding officers Gen. Winfield Scott and Gen. Philip Sheridan.

In the summer 1874 Custer led an exploration expedition with 1,200 troops and scientists into the Indian reservation lands of the Black Hills. Horatio Ross, one of Custer's two miners, discovered gold at French Creek. In his August 15th report Custer stated, "On some of the water-courses almost every pan-full of earth produced gold in small, but paying amounts. . . .It has not required an expert to find gold in the Black Hills, as even men without former experience in mining have discovered it at an expense of little tome or labor." Custer's reports were published not only in local but also national journals such as *Harper's Weekly*, and the news sparked a frenzied gold rush.

Much of the excitement had to do with the previous year's terrible economic depression, but some of it came from the belief that this might be the last great placer deposits to be found in America. John Gordon led the first group of prospectors into Indian land, the land where the Sioux gods lived, the *Pa Sapa* or sacred hills. Gordon's party evaded soldiers and Indians alike and reached the gold laden hills a few days before Christmas. These were hardy men who panned for gold in sub-zero weather. When April arrived and the snows melted, they were finally located and evicted from the Indian land. When Gordon's party reached the first settlements they spread the news of the gold finds. Soon the Army found that trying to keep out the gold rushers was like trying to stop the rising ocean tide with a broom.

Towns such as Deadwood, Central City, Custer, Lead and Rapid City grew up almost overnight between 1875 and 1876. As America was preparing to celebrate its centennial, another gold rush was drawing Argonauts like a magnet. There were the usual first timers along with veterans from all the other Western gold rushes. There were forty niners, miners from the Fraser River, miners who worked the Comstock and men who saw Pike's Peak. This was the point in time when the placer mining experiences of the past 27 years came together. A popular gold rushers song in 1876 sums it up:

You're looking now on old Tom Moore, a relic of by gone days;
A Bummer, too, they call me now,
But what care I for praise?
For my heart is filled with grief and woe,
And oft I do repine
For the days of old,
The days of gold,
The days of Forty-Nine.

Gen. George Cook was to become the finest Indian fighter in the U.S. Army. After two years he managed to pacify the Apaches in Arizona in 1873. In 1875 he took charge of the Department of the Platte. His new responsibilities included keeping the gold rushers out of the Black Hills until a new treaty was negotiated with the Sioux. The Sioux were less than thrilled with the prospect of giving up sacred land. Offers of cash payment did not move them. In an effort to show the federal government's good faith, Gen. Cook called a few meetings with the miners and told them that the government guaranteed their claims for 40 days after the Indians cleared out. This tactic worked in getting the miners to leave the Black Hills on more than one occasion, but it wasn't enough.

Red Cloud, who led the Sioux against the army in the battle for the Bozeman Trail, chose not fight this war. The Sioux warriors came under the control of Chief Crazy Horse and Sitting Buffalo, known to the Americans as Sitting Bull. The divisions among the Sioux after the breakdown of negotiations gave rise to an unusually large seasonal scattering in the fall of 1875. It was common for the Indians to find comfortable places to winter and this year many segments of the Sioux had much to discuss among themselves, away from the reservation and the prying

◄ *George Armstrong Custer finished last in his class in West Point but first in the imagination of America. His cavalry exploits during the Civil War and in the Indian Wars of Plains earned praise and status of legendary hero.*

◀
It was Custer's 1874 scientific expedition into the Black Hills that started the gold rush. Little did the colonel suspect that subsequent events would also lead to his death.

▲
After watching land that was granted to the Sioux by the U.S. government become overrun with gold rushers, Sitting Bull went to war and won his greatest victory, the Battle of Little Bighorn.

interests of the army. When a directive was issued by the government stating that all Indians had to return to their reservations by the end of January 1876, many in their isolated winter camps never got the word. Also, some did know about the directive but chose to ignore it. By spring those who did not return were considered hostile, and the Army made preparations to fight them.

The first encounter in June 1876 took place at Rosebud Creek in southern Montana. Gen. George Cook finally made contact with renegades and was attacked by 1,000 Indians. He had an equal number troops, but as the battle continued it became apparent to Cook that his men were outnumbered. It was as if for every Indian shot three more replaced him. Cook retreated to await reinforcements and the Sioux, together with their allies, the Cheyenne, crossed Rosebud Creek, rode over the ridge and traveled west to the next river, Little Bighorn.

The counterattack was commanded by Gen. Alfred H. Terry. His soldiers approached the Bighorn Mountains from the east, while a force under Custer's command was sent to swing south and come up on the rear of the Sioux and Cheyenne encampment. The strategy was a hammer and anvil attack. Custer was to wait until Terry's force was in place and then begin his surprise assault by using mobile cavalry to drive the retreating Indians into the general's slower, more numerous foot soldiers. It was expected that the Indians would be smashed much like a piece of steel that is laying on an anvil and being struck by a sledge hammer.

But Custer wanted to win the day on his own. He was used to taking chances and being right. To the victor goes the glory. Instead of waiting for Terry's troops to reach their positions, Custer split his forces into three commands and launched his own attack on the morning of June 25. Like Cook, he had no idea how many Indians were at the Little Bighorn encampment. Crazy Horse, shouting, "Today is a good to fight, today is a good day to die," reacted quickly and cut off the leading troops that were charging the camp. The central unit under Custer's control was speedily surrounded on a hill, and all 266 men and officers were killed. Maj. Marcus A. Reno and Capt. Frederick W. Benteen were commanding the two other units and they were able to re-

▶ *Some of the early mining towns in the Black Hills were little more than a couple of shacks. No one knew how long they would be able to pan for gold before the Army kicked them off of the Sioux land.*

treat to defensive positions. Gen. Terry and his troops arrived in time to save them. The whole battle took a few hours, but in the end the Sioux and Cheyenne victory signaled their doom.

When news of "Custer's Last Stand" reached the East on the morning of July 5 it invoked outrage in a nation that had just finished celebrating its centennial. It was a disgraceful defeat at the hands of what was thought to be a band of savages. A nation of 40 million wanted retribution and there was no limit to the number of soldiers it was willing to send in the effort. Crazy Horse was captured and later killed "attempting to escape." Sitting Bull and a band of Sioux escaped to Canada. Those Indians who survived the wars found a life of anguish waiting for them on the reservation. Three months later the Sioux formally ceded the Black Hills to the United States and the miners were free to hunt for their treasure. When all is said and done, Custer died for gold not glory.

Mining in the Black Hills started out as all the gold rushes that came before it did, with simple placer operations. By the end of 1876 lode claims were being filed. One such claim was found by Moses Manuel on April 6, 1876. He and his other three partners called the mine the Homestake. Shortly after filing the claim the four partners found a 200 pound nugget of quartz gold. It was the largest one ever found at the Homestake.

Events at the mine unfolded quickly. In Lead L.D. Kellogg, an agent of George Hearst, James Haggin and Lloyd Tevis, urged them to purchase the Homestake. The trio trusted Kellogg's assessment and let him negotiate the deal. For Manuel and his partners 1877 was a good year. They sold the Homestake for $100,000. For Hearst and his partners the following years were much better. The Homestake became one of the richest gold quartz mines in the world. It was estimated in 1931 that $230,000,000 in gold was mined from the Homestake.

After the purchase of the Homestake by the famous San Franciscan trio other quartz claims were bought by investors and industrial technology rapidly moved in. First came the mining engineers and heavy machinery, followed by the building of stamp mills and amalgamating plants. This created a need for professional un-

◄ *Miners were always in need of supplies, and some men preferred the surer income from hauling goods than the riskier business of mining. Freight wagons such as this one crossed the roadless prairies to towns, Rockerville and Keystone, for example, in search of profits.*

▲

After the Battle of Little Bighorn the Cavalry's sole interest was in defeating the Sioux and protecting the incoming miners and settlers.

▶

Mining activity in the midst of a growing town occurred often during the gold rushes, and Deadwood was no exception. In this photo men are mining for gold while the town is being built around them.

derground miners whose jobs relied on the support of mechanics, bookkeepers and equipment operators, many of whom brought their families. The area grew, bringing the railroads to its doors in 1886.

Four miles from the Homestake Mine grew the town of Deadwood, located in Deadwood Gulch. If the placer deposits of the area attracted gold rushers, Deadwood lured legends. As did many a gold rush town, Deadwood started out as a collection of rough lean-tos and flimsy tents. Within a short time log cabins made their appearance, followed by hand sawed pine board structures. The town itself was laid out on April 26, 1876 and construction began immediately. Main Street was an odd shoestring affair that meandered along the bottom of the canyon and was occasionally obstructed by the sluices and diggings of the miners. By the summer Deadwood had a population of 25,000, which included every conceivable type of humanity that could be packed into the hodgepodge city stretching lengthwise along the gulch. Deadwood was as much a part of the legendary West as Custer's Last Stand.

For some unknown reason many of the most well-known personalities of the West visited, lived and in some cases even died in Deadwood. They were Wild Bill Hickock, Buffalo Bill Coty, Calamity Jane, Wyatt Earp, Doc Halliday, and Bat Masterson, as well as outlaw Sam Bass. Perhaps it was the vanishing freewheeling spirit of the frontier that brought these people there in the first place.

Wild Bill Hickock was the one famous Western personality whose murder in Deadwood became a legend. Born James Butler Hickock on May 27, 1837, he lived in Homer, Illinois until 1856 when he journeyed to Kansas to help with the free state movement. Abolitionist sentiment ran strong in the Hickock family. Their secret basement, where they hid runaway slaves, was part of the Underground Railroad.

Hickock's reputation as the fastest gun alive began in 1861. The young Hickock, who was not known as "Wild Bill" yet, was working at the pony express service station when David Colbert McCanles, the owner of the property on which the station stood, called on the new owner of the business. McCanles, with two friends, had come to collect the back rent that was rightfully his. An argument broke out and Hickock shot McCanles, his two friends and the owner of the sta-

tion before anyone could draw their guns. All this was said to be over a woman who had spurned Hickock's advances and favored McCanles instead.

Another more popular point of view describes the incident differently. McCanles sold the land and wanted some of the money that he felt was still owed to him. When he showed up with three of his employees an argument broke out because McCanles decided to take some horses in lieu of payment. Hickock tried to stop them and the would be horse thieves went for their guns. They were all gunned down within four seconds, one second per bullet per man. In either case, whichever story is true, Hickock was acquitted of all charges and headed for Missouri to join the Union forces in the Civil War.

During the war Hickock made a name for himself as a spy, a scout and a sharpshooter in Missouri. It was his wartime adventures that gave

▲
In 1876 America was celebrating its centennial and Deadwood City was busy attracting the legends of the West. It became the place to be if your name was Hickock, Holliday or Earp.

him the reputation as the fastest gun in the West. After the war, when Hickock was out of work, he supported himself by gambling. In 1866 he was made deputy marshall of Fort Riley, Kansas. Later, as a U.S. marshall for the violent cattle towns of Hays City in 1869 and Abliene in 1871, and as an Indian fighter with Gen. Hancock, Gen. Sheridan and Col. Custer, his claim to fame as a Western legend reached its highest point. Hickock attempted a brief career in show business touring with Buffalo Bill Coty in the play *Scouts of the Prairies*, and was part of Buffalo Bill's Wild West Show until 1873. There is some speculation that he may have fathered a child with Calamity Jane during this period, although many of his biographers deny it. During the last few years of his life Wild Bill returned to gambling. It was this profession that brought him to Deadwood.

When he arrived in Deadwood Hickock was married, not to his friend Calamity Jane but to Agnes Lake Thatcher. Everyone knew who he was and some respected and feared him. Wild Bill had an uneasy feeling that his presence seemed to stir up a few of the more violent men in town. "Those fellows over across the creek have laid it out to kill me and maybe they're going to do or they ain't. Anyway, I don't stir out of here unless I'm carried out." It may have been this nagging murder suspicion that made Wild Bill insist on changing his seat in a poker game. Thus, he didn't have his back to the door on that warm evening of August 2, 1876. The other players at the table thought he was just being dramatic. At a few minutes past three Jack McCall walked behind Hickock and shouted "Damn you, take that!" as he shot Wild Bill in the back at point blank range. Hickock died instantly and slumped forward revealing his poker hand, two pairs, eights and aces, known forever afterwards as the Dead Man's Hand. McCall was later hung for the crime and Hickock lived on as a legend.

Deadwood had a few fitful years that lasted into the eighties. It may have been the stagecoaches traveling out of the Black Hills that gave rise to the cliche Western movie scene of the hold-up. It was 60 hours From Deadwood to Cheyenne, longer if the roads were muddy. Over these roads gold laden stagecoaches provided too tempting a target to be ignored. The gold from the placer sites was carried in ordinary steel strong boxes, similar to the ones seen in the movies. The bandit shot the lock off, opened the lid and collected the loot. Later, the boxes were bolted to floor of the coach, which only meant that all the victims had to stand around while the outlaws opened the box. The next attempt to foil the crooks was made by adding extra guards. Wyatt Earp was once hired to ride shotgun on one of these stagecoaches. In 1878 stagecoaches evolved into armored treasure coaches. One inch thick steel plates that could stop a rifle bullet

▲
Gambling saloons followed easy money during the gold rush days. It was in a saloon such as this that Wild Bill Hickock was killed while holding what came to be known as the dead man's hand of aces and eights.

CHAPTER 10

THE KLONDIKE

Alaska, situated in the extreme northwestern part of North America, is the largest state. It was thought that the name "Alaska" came from the Eskimos and meant "great country," however, although the name does come from an Eskimo word, *Alakshah*, it means "mainland." Alaska is bordered on the north by Arctic Ocean; on the west by Arctic Ocean, the Bering Strait and the Bering Sea; on south by the Pacific Ocean and on the east by the Canadian Yukon Territory and British Columbia. The three geographical regions of Alaska consist of mountainous southern Alaska, the low, broad area of the Interior and the Arctic Alaska, which is in the north and separated from the Interior by the Brooks Range.

A popular misconception about Alaska is the freezing Arctic temperature. In fact Alaska has wide variations in climate and the famous glaciers can be found in the mountains as far south as the Canadian border. The southern part of Alaska has an oceanic climate with no great extremes of temperature.

Alaska was discovered in 1741 by Vitus Jonassen Bering on his second voyage to determine whether Siberia and North America were linked. Bering's ship was wrecked on Bering Island and he died, but his crew returned to Russia with sea otter pelts. This peaked Russian interest in Alaska and for the next 60 years the Alaskan coast attracted a hoard of Russian fur traders and trappers. These men had only one quest in mind, to get the furs any way possible. This included stealing when the Aleuts would not sell to the Russians. Unfortunately, the Russians did not stop there. They kidnapped and enslaved Aleut women on occasion, sometimes massacring Aleut men in the process.

At the beginning of the 18th century things settled down to a more businesslike arrangement with the formation of the Russian American Company. It was a defensive move designed to block out British and Spanish interests in the area. Whaling and fishing were important, and the United States became a commercial rival. In 1821 Russia issued a decree that forbade other nations from harvesting the waters off the coast of Alaska. Unfortunately for the Russian interests, the Alaskan colony was dependent on foreign trade for its survival. The decree was unenforceable. With the outbreak of the Crimean War in 1853, Russia found that her Pacific colonies were in a vulnerable position. By 1857 the Russian minister to Washington began hinting that the Czar would be willing to sell Alaska to the United States. Negotiations began in 1859, but the Civil War delayed any further considerations of the sale. Finally, in 1867, a treaty was formed and the United States purchased Alaska for $7,200,000. Many Americans thought it was too much money for a large tract of useless land and

SCALES

they called it "Seward's Folly" after William Henry Seward, the man who negotiated the final deal.

According to Jack London gold was discovered as early as 1804 by the Russians in Alaska. London's story in the Atlantic Monthly Magazine in July 1903 stated that a Russian governor at Sitka, after being shown about 50 small nuggets, discouraged anyone from talking about the find. He was sure that the news would encourage a flood of gold rushers and spell the end of his livelihood.

Two decades later the Fraser River gold rush miners searched the various creeks of southern Alaska hoping to be the first to find gold in virgin territory. In 1880 Joe Juneau and Dick Harris discovered rich placer deposits in Silver Bow Basin on the southeastern coast. By 1881 the gold rush town of Juneau made its appearance on the map along with thousands of miners who inhabited the town. The Juneau discovery was only mildly productive. Another nearby quartz outcropping discovered by Pierre (French Pete) Erussard became one the top 10 productive gold mines in the 19th century world. Once again it was the story of the miner selling out to someone else for a fraction of the gold site's worth. In this case John Treadwell, a builder and contractor from San Francisco, bought the mine. With the aid of San Franciscan investors he formed the Treadwell Company to mine the gold.

It was not a placer project, but deep lode quartz mining. Treadwell could always find miners to work in his mine. He paid above average wages and was considered progressive for his time. But the miners never stayed long. They were placer folk looking for the huge gold find that would set them up for life. By the end of the 19th century the Juneau mines had unearthed over $17,000,000 in gold.

Although there were numerous small gold finds before 1892, no event had a greater impact on the later Yukon and Klondike gold rushes than the opening up of the Chilkoot Pass. In 1879 a group of 20 prospectors attempted to cross the Chilkoot Pass into Alaska, but they were prevented from climbing the trail by the Indians. The Indians had the same fears as the Russians, that their Yukon fur trade might be destroyed if the pass was opened to American gold hunters.

◄ *The Scales, Chilkoot Pass was in Canada, and the Mounties made the gold rushers carry enough provisions to last a season. This entailed climbing the pass for 10 days, often three or four times a day, until the required amount of supplies was reached.*

Help came from an unexpected source, the commander of the S.S. *Jamestown*. The captain sent his naval lieutenant to the pass. There the Indians saw first hand how a Gatling gun worked. It must have been an impressive demonstration indeed, because on May 29, 1880 the Chilkoot Pass was officially opened to white men for the first time.

In 1886 miners struck a rich placer field on a branch of the Forty Mile River, a region that straddled the Alaska-Yukon Territory border. Later, the first interior gold town developed at Forty Mile in Canada, a mere 25 miles from the border. It was not a spectacularly rich gold site, but there was enough gold to keep several hundred miners busy for six years. When the placers looked as if they would soon be exhausted, the mining activity shifted 240 miles away to the Birch Creek area. Out of this small rush grew Circle City.

Circle City was the first gold rush town in Alaska's interior. It also was the first to look and feel like a gold rush town with saloons, dance halls, an opera house, prostitutes and regular steamboat service to the Yukon. It was the Circle City miners who gave rise to the myth of the well read gold rusher. It began with a U.S. senator who visited the town to see how the miners lived. As he told the story back in Washington, he was in a miner's cabin when he noticed that it contained a small library. There were books on science, technology, Shakespeare and six volumes of Gibbon. "Andy," the senator asked pointing to the books, "who left them here?" "Who left them?" answered the miner angrily, "why no one. I've had 'em for two or three years. Take 'em everywhere, and read 'em nearly every night when I get time. I'll bet I know more about Caesar, Hadrian, Attila, and all the others than you do, or most anyone else."

Late in 1896 gold was discovered on the Klondike, a tributary of the upper Yukon in Canada. It was originally found by an experienced prospector, Bob Henderson, at Rabbit Creek, later known as Bonanza Creek. Henderson's biggest problem was that he hated Indians, and the first people he told about the discovery were a Siwash Indian, Skookum Jim, his white brother-in-law, George Carmack, and another relative of Skookum Jim's, Tagish Charley. Actually, the way Henderson told the story, he only told Carmack about the find, not realizing that the man had a Siwash wife. Carmack's version was an elaborate story full of details on how he outsmarted Henderson. Skookum Jim's tale was very matter of fact. The trio had been out hunt-

▲
No one can properly describe the hardships endured by early gold rushers who travelled the Chilkoot or White passes. Without necessary supplies the hungry and exhausted later day Argonauts, along with their pack animals, found themselves in a barren, rocky hell, which was later called Dead Horse Trail. The gold rushers beat their horses to keep them moving, because once the animals stopped they would fall to the ground and die. Jack London wrote: "Their hearts turned to stone–those that did not break. And they became beasts, the men on the Dead Horse Trail."

ing and Carmack fell asleep under a tree. The Siwash took his frying pan down to the creek to wash it and got this notion that it might be an interesting place to pan. In no time at all Skookum Jim had panned about $4 worth of gold. This was considered good in the Yukon where 10¢ a pan was the norm.

At first many of the old hands at Circle City and Forty Mile refused to believe any version of the gold strike. They had all heard it before. Shopkeeper Joseph LaDue, the man who had grubstaked Henderson, heard the news and immediately laid out the townsite for Dawson City. He knew how money was made. Soon large quantities of gold started showing up in Circle City and the miners found their fate. The gold rush was on.

This time fate was kind to most of the original finders. Tagish Charley sold out in 1901 and opened a restaurant in Carcross. His fortune was his ruin. As owner of a fine establishment he allowed himself to drink in excess. One drunken night he fell off of a bridge and drowned. Skookum

Jim earned $90,000 year from his share and kept prospecting. Some say he died from overwork in 1916. Bob Carmack became a millionaire, left his wife, remarried and settled down in Vancouver. At the time of his death in 1922 he owned another mine in California, an apartment house and a hotel. Henderson was recognized by the Canadian government as the rightful discoverer of the Klondike gold and given a $200 a month pension. He never made any other money from his claims and died a broken man.

It was another slow start to the rush because the rest of the world was even more skeptical than the miners in Circle City. By July 1897 there were only 4,000 gold rushers in the Klondike. The real excitement came when two steamers arrived in Seattle and San Francisco carrying gold valued at $1,500,000. That was proof enough for even the most hardheaded. Newspaper headlines screamed about "Inexhaustible Riches of the Northern El Dorado," and some hundred thousand men and women started out for the North to reap its riches. Dawson City was the goal. About 40,000 Argonauts made it there and another 10,000 ended up in Alaska.

Mining was done at the early California level of technology, with a few necessary cold weather twists. The bedrock where the gold lay was anywhere between 20 and 200 feet underground. In order to penetrate the permafrost, miners had to burn out a shaft with wood fires. The back-breaking process of thawing, digging out the dirt, and thawing again finally led to the bedrock. The gold bearing gravel then had to be hoisted out of the hole in buckets and stored in large piles until summer. With the warmer weather came the water needed to process the gravel in pans and sluices. It was not a lot of fun, but it was profitable. From 1897 to 1900 $50,000,000 in gold was mined in the Klondike.

With so many gold rushers in the region it was inevitable that gold would be discovered somewhere else. The Nome gold rush in Alaska soon followed. Prospectors discovered gold on Anvil Creek in 1898. In 1899 claim jumpers, in reality outlaw miners, sought to drive out many of the original miners. Tensions were high in Nome because fighting was the order of the day.

With the discovery of gold on Nome Beach thousands of men and women began prospecting the coastal sands in spite of the constant danger from storms coming in from the Bering Sea. But each storm that came would uncover more gold so more gold rushers wanted to journey to the Nome Beaches.

CLOTHING
GENERAL OUTFITTIN
BOOTS & SHOES
SHOES
DR BOURKE
GEO. NOBLE
MANAGE

By 1900, even though the gold rushers were mining over $2,000,000 a year, many were leaving Nome just to stay alive. Into this confused situation stepped district judge Arthur H. Noyes and his friend Alexander McKenzie. The conflicting mining claims provided Noyes with the opportunity to appoint McKenzie receiver and operator of the disputed sites. The two men defied a court order to make restitution of all properties. Finally, in 1901, Noyes was removed from office bringing an end to this audacious scheme.

Mining activity spread throughout the Seward Peninsula, but the next major gold discoveries were made deep in the interior at Tanana Valley in 1902. Nearby the town of Fairbanks grew to become the largest center of population in the territory. As the first major town to be found away from the coast, Fairbanks had an enormous impact on the development of the Alaskan interior.

In 1904 industrialized mining came to Alaska in the form of dredge mining. Dredges were able to scoop out bedrock gravel and wash it down automatically. Steam or hot water were driven into the ground to defrost the earth and hydraulic hoses washed away the overlay of soil covering the bedrock. The first large dredge to be used in mining operations was in Nome. The 20th century had come to Alaska and the great gold rushes of North America were over.

◀

Dawson was unlike other gold rush towns it was relatively quiet. Movies have shown the women of Dawson dressed in clothing meant to allure the miners, but in fact the moral tone of the town took its cue from Queen Victoria, the Salvation Army and the many nuns who came to Dawson. Indecent exposure was out and the rustle of silk was in. The Mounties strictly enforced the laws and moral codes. Many a gold rusher remembered dull long winters where men caught cabin fever for the lack entertainment and social life.

▲
Only the simplest machines could be used all year round on Nome Beach. Heavier contraptions would be destroyed by storms.

BIBLIOGRAPHY

Adams, Alexander. *The Disputed Lands*. New York: G. P. Putnam's Sons. 1981

Angier, Bradford. *Looking for Gold: The Modern Prospector's Handbook*. Harrisburg, Pa: Stackpole Books 1980

Ballentine, Verne H. *How and Where to Prospect for Gold*. Blue Ridge Summit, Pa: Tab Books Inc. 1981

Billington, Ray Allen. *The Far Western Frontier*. New York: Harper & Row, Publishers. 1956

Brandon, William. *Indians*. New York: American Heritage Press, Inc. 1985

Burns, E. Bradford. *A History of Brazil*. New York: Columbia University Press. 1980

Chidsey, Donald Barr. *The California Gold Rush*. New York: Crown Publishers 1968

Coleman, Kenneth, I. *A History of Georgia*. Athens, Georgia: University of Georgia Press. 1977

Eaton, Clement. *A History of the Old South*. New York: Macmillan Publishing Co. 1975

Flanagan, Mike. *Out West*. New York: Harry N. Abrams, Inc. 1987

Hawn, Emily. *Love of Gold*. New York: Harper & Row, Publishers. 1980

Herring, Hubert. *A History of Latin America*, Third Edition. New York: Alfred A. Knopf. 1968

Holliday, J. S. *The World Rushed In*. New York: Simon and Schuster. 1981

Horan, James David. *The Gunfighters: The Authentic Wild West*. New York: Crown Publishers Inc. 1976

Jackson, Donald Dale. *Gold Dust*. New York: Alfred A. Knopf. 1980

Lacour-Gayet, Robert. *Everyday Life in the United States Before the Civil War 1830-1860*. New York: Frederick Ungar Publishing Co. 1969

Lamar, Howard Roberts. *Dakota Territory 1861-1889*. New Haven, Conn: Yale University Press. 1956

May, Robin, *The Gold Rushes*. London, Great Britian: William Luscombe Ltd. 1977

McGloin, John Bernard. *San Francisco, The Story of a City*. San Rafael, Calif: 1978

Morrell, W. P. *The Gold Rushes*. New York: Macmillan Publishing Co. 1941

Petralia, Joesph F. *Gold, Gold! A Beginner's Handbook*. British Columbia, Canada: Hancock House Publishers Ltd. 1980

Quiett, Glenn Chesney. *Pay Dirt*. New York: D. Appleton-Century Co. Inc. 1936

Schlissel, Lillian. *Women's Diaries of the Westward Journey*. New York: Schocken Books. 1982

Smith, Duane A. *Rocky Mountain Mining Camps*. Bloomingon, Indiana: Indiana University Press. 1963

INDEX